Donkey

Animal
Series editor: Jonathan Burt

Already published

Crow
Boria Sax

Ant
Charlotte Sleigh

Tortoise
Peter Young

Cockroach
Marion Copeland

Dog
Susan McHugh

Oyster
Rebecca Stott

Bear
Robert E. Bieder

Bee
Claire Preston

Rat
Jonathan Burt

Snake
Drake Stutesman

Falcon
Helen Macdonald

Whale
Joe Roman

Parrot
Paul Carter

Tiger
Susie Green

Salmon
Peter Coates

Fox
Martin Wallen

Fly
Steven Connor

Cat
Katharine M. Rogers

Peacock
Christine E. Jackson

Cow
Hannah Velten

Swan
Peter Young

Shark
Dean Crawford

Rhinoceros
Kelly Enright

Moose
Kevin Jackson

Duck
Victoria de Rijke

Horse
Elaine Walker

Elephant
Daniel Wylie

Eel
Richard Schweid

Ape
John Sorenson

Pigeon
Barbara Allen

Owl
Desmond Morris

Snail
Peter Williams

Hare
Simon Carnell

Penguin
Stephen Martin

Lion
Deidre Jackson

Camel
Robert Irwin

Pig
Brett Mizelle

Vulture
Thom van Dooren

Lobster
Richard J. King

Donkey

Jill Bough

REAKTION BOOKS

Published by
REAKTION BOOKS LTD
33 Great Sutton Street
London EC1V 0DX, UK
www.reaktionbooks.co.uk

First published 2011

Printed and bound in China by Eurasia

British Library Cataloguing in Publication Data
Bough, Jill
Donkey. – (Animal)
1. Donkeys 2. Donkeys – History.
3. Donkeys – Symbolic aspects.
I. Title II. Series
636.1'82-DC22

ISBN 978 1 86189 803 6

Contents

Introduction

> He can live without man. But man can scarcely do without the labour, the sacrifice, the suffering of the donkey . . . that has accompanied man since the dawn of time, in all weathers, humbly and patiently serving the most brutal of all animals.[1]

Donkeys are commonplace. They live in most areas of the world alongside humans, an integral presence in many cultures. Even if they are no longer useful to human endeavours in much of the developed world, they are still pulling carts in Africa, bearing heavy loads in India, carrying tourists in Greece and taking children for rides along British beaches. Considering how long they have been domesticated and how valuable they have been in human history, we know remarkably little of their lives or their stories, or even of their welfare. It is not that they are unfamiliar; it is that they are generally considered beneath notice. As they crossed the world in the service of their human masters, donkeys have been among the most used, and abused, animals in history.

Donkeys have served humans, largely as beasts of burden, since the time of their domestication, possibly 10,000 years ago.[2] They carried crops and produce, hauled rock and timber, moved provisions and merchandise, transported food and equipment for armies and turned heavy millstones to produce flour. Even today, donkeys continue to make an important contribution to some economies in the developing world, such as those of Ethiopia, India and Pakistan, and are indispensable to the survival of some of the poorest societies in the most inhospitable places – for example, in salt collection and transportation in Africa's Danakil Desert.[3]

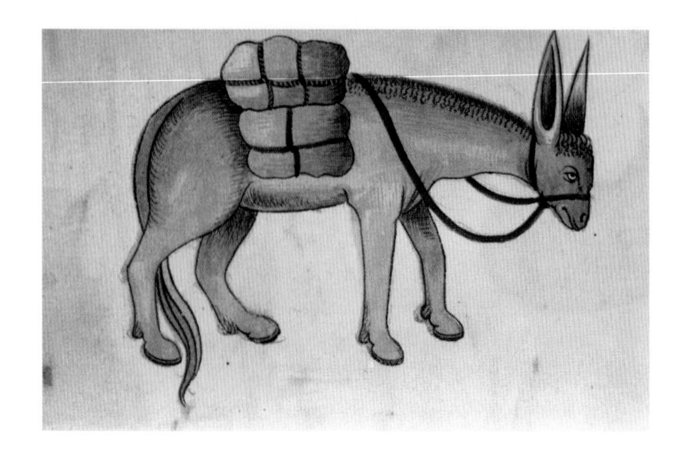
Illustration of a donkey with a pack-load, c. 1520s.

It is estimated that there are 41 million donkeys in the world today, 51 per cent in Asia, 28 per cent in Africa and 18 per cent in Central and South America. A report from the Food and Agriculture Organization (FAO) in 2006 revealed that donkey populations were continually diminishing globally despite the animal being the main form of transport in many parts of the developing world. Although ignorance and prejudice with regard to the value of donkeys remain in many areas of Africa today, mainly because of the implications of lack of progress because of the animals' lowly and backward status, experts in animal traction consider donkeys to be one of the best draught animals. They are reported to have the longest working life, to be able to work in the driest areas, to manage on the least food, to be the least disease-prone, to be able to work at variable speeds and to have a high learning ability by comparison to horses, mules and oxen.[4]

However, despite the service that donkeys have rendered to humans in all ages and societies, they have received little recognition in return. The injustices and indignities they have received

Donkeys labouring in the baking temperatures and brutal landscape of the salt plains of the Danakil Desert, Ethiopia.

are cause for reflection by various commentators. For instance, when Reverend J. P. Mahaffy delivered a paper on the Irish donkey in London in 1917, he speculated how this dignified animal 'should have been for centuries the emblem of stupidity and ridicule'. The paper was received cordially and with interest but also with much hilarity. He had cause to wonder why this hard-working animal aroused such scorn.[5]

This donkey in harness is powering a millstone for grinding dried red peppers, Binxian City, China.

Francisco de Goya, 'Tu que no puedes' from the *Caprichos* series, 1797–8.

However, I maintain that it arises from both what the actual animal, the donkey, is and does and, perhaps even more importantly, from how we choose to represent donkeys in human terms (or humans in donkey terms); the language used to describe them usually involving demeaning comparisons, as will be explained in the following paragraphs. The words 'donkey' and 'ass' both refer to the domesticated animal descended from the wild asses of Africa: their origins, history and physiological attributes will be considered in chapter One. Donkeys are hardy and resilient animals that can work tirelessly with little maintenance. Although usually slow paced, they are steady and sure-footed. They are also strong and sturdy and can carry heavy burdens relative to their size. They are, for example, often expected to carry loads equal to two-thirds or more of their body weight, as has been recorded in Mexico, Peru, Ecuador, Venezuela, Greece, Tunisia, Spain and throughout most of Africa. Although known as good workers, donkeys have a strong sense of self-preservation, which has influenced their reputation for stubbornness.

The origins of the name 'donkey' are not clear, and have changed over time and with context. In the language of their early owners, the Semites, donkeys were called *anah* and in

A donkey buried in a huge hay bail on a road in rural Ethiopia.

Latin *asinus* (which later became *ane* in French). The etymology confirms that donkeys were established throughout the Mediterranean, Levantine and Anatolian regions long before Indo-European horse users arrived. The derivation of the word 'donkey' to describe the domesticated ass is not really known but it is suggested that it could be from 'dun', or dull grey-brown coloured, and the form perhaps influenced by the word monkey. A male donkey is called a jack and a female a jenny or jennet. In parts of the United States where Latin American culture persists, *burro* is more commonly used than donkey. In Britain, as in many other countries, local names for donkeys have evolved in different dialects, such as 'moke', 'neddy' and 'cuddy'. However, the largest variety of names is to be found in Hebrew, as donkeys were so important in the lives of Semitic peoples for so long. Perhaps the best known of these is *hamor*. One of the rulers of the Semites was named Hamor (who governed the city of Schechem in the seventeenth century BC), so to call someone that name was certainly not considered an insult in Hebrew culture. The same could not be said of calling someone 'ass' or 'donkey' in English-speaking cultures.

In part, the reason why donkeys have become figures of ridicule lies in the ways in which humans represent them, both in the language we use to describe them, and in the phrases we apply to human behaviour that derives from depictions of donkeys. The Arabic *himar*, for example, is used to refer to a person of limited intelligence, lacking in skill or unsuccessful. The term 'asinine' to describe a person implies foolishness, someone lacking in intelligence, even retarded, despite the fact that this is untrue of donkeys themselves. Other words, phrases and sayings have become part of common usage in various cultures to denigrate humans by comparing them to donkeys. In both football and poker in the UK, for example, a player called a donkey is one who lacks skill. In Australia a so-called donkey vote occurs when an elector simply numbers the ballot paper from top to bottom (or bottom to top) without regard to the logic of the preference allocation; since voting is compulsory in Australia 'donkey votes' are usually considered to be a sign of ignorance or apathy. 'Donkey lick' is another Australian idiom that means to defeat conclusively, as in a horse race. Someone who can 'talk the hind legs off a donkey' never stops mindlessly prattling on, paying no heed to anyone or anything else.

More complex are the meanings associated with the word 'ass' because of the semantic associations that developed between the animal and human buttocks.[6] If a person is foolish, they are said to be 'making an ass of themselves' or 'not knowing their ass from their elbow'; while to refer to someone as a 'dumbass' or a 'jackass' has more deeply insulting connotations, implying complete stupidity, even vulgarity. A mean person is a 'tight-ass', while a sly one 'covers their ass' and a show-off is 'a smart ass'. However, there are also a number of positive connotations: of strength, as in 'to kick ass', and of hard work, as in 'to work your ass off'. 'To be hung like a donkey' could be considered a compliment by some as it refers to the jack's large, relative to his size, penis. As we will see in chapter Two, this aspect of the donkey's physiology resulted in its symbolic link with ancient fertility gods, an association that in time led to the denigration of the donkey.

Donkeys also came to be associated with the poor, oppressed or marginalized in society. Although in his history of domesticated animals Frederick Zeuner finds the donkey 'indisputably one of the most useful animals', he also finds it to be 'despised almost everywhere'. He suggests that 'its stolid temperament has annoyed its master since time immemorial . . . The patience of the ass is likened to the slave.'[7] Zeuner echoes the thoughts of

The low status of the donkey is affirmed by this illustration from *Uncle Tom's Cabin*; and is ridiculed as a comic outcast.

the eighteenth-century French naturalist Georges-Louis Leclerc Comte de Buffon, writing about the ass in his influential *Histoire naturelle* (1749–88). Commenting on the poor treatment of the ass, he posed these questions:

> Why, then, should an animal so good, so patient, so temperate, and so useful, be treated with the most sovereign contempt? Do men despise, even in the brute-creation, those who serve them best, and at the least expense?[8]

As donkeys made their way from Africa to Europe, their fortunes changed as they lost status and were physically abused and neglected. Their symbolic role in religion and mythology also

A supposedly scientific depiction of a donkey, from the Comte de Buffon's *Histoire naturelle* (1800), indirectly comments on the poor treatment of donkeys.

LAWYER MARKS-
UNCLE TOM'S CABIN

Donkeys are the most common means of transport in rural Ethiopia. Here a Fur girl (the Fur are the largest ethnic group in the Darfur region of western Sudan) is watering donkeys at one of the few wells in Darfur.

led to their denigration. Ironically, their lowly status in Africa today reflects those changed attitudes.

The identification of donkeys as humble has deemed them suitable for women's use in many societies; they have none of the 'masculine' characteristics associated with wealth, power and status. Perhaps because they are easy to manage and train, however, there are fewer gender restrictions associated with owning donkeys. In most countries they are used both by men and women, with many women's tasks (hauling water and firewood, for example) being alleviated. Two Ethiopian proverbs claim that 'Donkeys and women can carry whatever reaches their backs' and 'A woman without a donkey is a donkey herself'. Donkeys have reduced the domestic transport burden for many rural women and have also created employment for them.[9] The donkey's symbolic significance in matters of gender has long been recognized in many societies. In those based on a patriarchal ethos especially, donkeys have been used as objects of derision, associated with cuckolded husbands. In post-Renaissance France and England, society ridiculed and humiliated husbands who were dominated by their wives. In France a 'battered' husband was trotted around town riding a donkey backwards while holding its tail. In England, 'abused' husbands were strapped to a cart drawn by a donkey and paraded around town to public redicule.[10]

However, it is in comparison with horses that donkeys have suffered most. Buffon encapsulated this problem in his *Histoire naturelle*. In this inappropriate comparison, the donkey is invariably found wanting – and is treated accordingly.

> The horse we educate with great care; we dress, attend, instruct, and exercise him: While the poor ass, abandoned to the brutality of the meanest servants, or to the malicious abuse of children, instead of acquiring, is rendered more stupid and indocile, by the education he receives. If he had not a great stock of good qualities, they would necessarily be obliterated by the manner in which he is treated . . . he is neglected and despised. It is comparison alone that degrades him. We view and judge him, not as he is, but in comparison with the horse . . . In his disposition, the ass is equally humble, patient, and tranquil, as the horse is proud, ardent, and impetuous.[11]

To be mounted on a horse confers power and authority on the rider; to ride a lowly donkey implies penury, lack of status, even stupidity. In the ancient world, the horse represented freedom,

'Journey of a Modern Hero', 1814, hand-coloured etching.

power and beauty, qualities that humans valued in the beast and desired for themselves. In Homer's *Iliad*, for example, two comparisons illustrate this point. The Trojan prince, Paris, is compared to a horse who, freed from its stable, gallops across the open plain, proud and strong, with mane flowing and head held high.[12] In another example, the Greek hero, Ajax, in his fight against the Trojans, is compared to a donkey in a field that will not lift his head from eating grain, despite being beaten with sticks. He is greedy, stubborn and mundane, unlike the horse that is motivated by the finer attributes of freedom and mastery.[13]

Such imagery continues to this day and it is invariably these social constructions that most affect how an animal is valued and treated. However, as we will see, these constructions can vary considerably according to context, and alter over time. Robin Borwick, for instance, refutes the donkey's reputation in comparison to the horse's: 'It is my opinion, and I am quite prepared to argue the hind leg off a horse about this, that a donkey's intelligence is greatly superior to that of the horse.'[14] As well as arguing for the intelligence of donkeys, enthusiasts describe them as hard-working, loyal, friendly, playful, gentle and eager to learn, while others write of the nobility, humility and wisdom of donkeys. One observer in 1935, moved by the sight of a line of donkeys carrying their loads, waxed even more lyrical in an account in the *New Mexico Historical Review*:

> We see plodding patiently along the country byway little trains of little burros, each bearing on his diminutive back a load much bigger than himself, but suffering his trials with so much patience and uncomplaining good humour that the conviction flashes upon my mind that

The old world order meets the new as donkeys carry computers on the Greek island of Hydra, where the only form of transport is by donkey and mule.

> each burro is now the place of transmigration of the soul of some ancient stoic philosopher; a conviction which impels me to touch my hat to a burro every time I meet one.[15]

Donkeys themselves have not changed, only human conceptions of them. This becomes apparent in chapter Three, which explores the donkey's roles in three colonial settings. Chapter Four explores the role of donkeys and mules in various theatres of war, while chapter Five considers a variety of representations of donkeys in literature and art, which continue both to reflect and to inform our perceptions and treatment of them.

1 *Equus asinas*: Origins, Domestication, Breeds and Characteristics

Miniature donkeys possess the affectionate nature of a Newfoundland, the resignation of a cow, the durability of a mule, the courage of a tiger, and the intellectual capability only slightly inferior to man's.[1]

Donkeys are relatives of the horse (*Equus*), belonging to the same order of odd-toed ungulates. The horse family includes the wild horse (*Equus caballus*), wild ass (*Equus hemionus*), zebra (*Equus quagga*) and the domestic donkey (*Equus asinus*). It is possible for all these species to interbreed in the wild and there have been many types of equid hybrid; under conditions of domestication, it is possible to produce hybrids between all equid species. However, the offspring of such interbreeding is nearly always sterile. The hybrid has the great advantage to humans of heterosis or 'hybrid vigour', which means that it is likely to be larger, have greater endurance and survive on poorer food than either of its parents. The first domestic hybrids were most probably crosses between wild onagers (*Equus hemionus*) and the early domestic donkey. However, the cross that has been most significant in human history is that between horses and donkeys: the mule.

The African wild ass (*Equus africanus*), the progenitor of the domestic donkey, is now critically endangered. These strong and graceful animals once roamed wild and free, galloping across the arid and rocky scrublands of Africa. Ideally suited to their semi-desert homes, they thrive in any dry, stony area where there is

A mule and a horse, from a 12th-century manuscript of the medical authority Sextus Placitus' fanciful descriptions of medicines derived from animals, *De virtutibus bestiarum in arte medicinae*, 4th century AD.

A mule from Sextus Placitus' *De virtutibus bestiarum in arte medicinae*.

scrub vegetation and where they are within two or three days of water. The beauty, speed and deportment of these wild animals excited much interest in European explorers, as well as the big game hunters of the nineteenth century. British explorer Samuel Baker described them while travelling in Africa in 1868:

> Those who have seen donkeys in their civilized state have no conception of the beauty of the wild and original animal,

The African wild ass, now critically endangered, is the ancestor of the domestic donkey.

> far from the passive and subdued appearance of the English ass, the animal in its native domain is the perfection of activity and courage; there is a high-bred tone in deportment, a high actioned step when it trots freely over the rocks and sand, with the speed of a horse as it gallops over the boundless desert.[2]

Of the two species of African wild ass, the Somali is taller, greyish in colour and is distinguished by strong dark stripes on its long legs. These asses still manage to survive in the wild in Somalia, Eritrea and Ethiopia despite a 90 per cent reduction in their range in the last twenty years. The sturdier Nubian wild ass, depicted as a hunted animal in Egyptian art and known for its stamina, is generally believed to be the ancestor of the domestic donkey. It is fawn in colour and, as well as an 'eel' marking running down its back, has a distinct shoulder stripe running from the withers down to the top of the leg, inherited by the domestic donkey. The ears are typically longer than those

Although endangered, a few Somali asses survive in the wild in Africa; but this one resides at St Louis Zoo, Missouri. Note the distinctive leg stripes.

of other wild asses and the bray is identical to that of the donkey. Nubian wild asses have only been infrequently seen since the 1970s and are therefore considered possibly extinct, with only a few remaining in captivity. Severe drought, devastating wars, loss of habitat and competition with domestic livestock for pasture and water are the main reasons for their drastically reduced numbers. Their capture for domestication over centuries has also contributed, as have interbreeding between wild and domestic animals and being hunted for food and for traditional medicine in both Ethiopia and Somalia.[3] These critically endangered African wild asses exist today only in small numbers in the zoos and wildlife preserves of eastern Africa. Ongoing research by leading academics in the field, concerning the distribution of wild asses and the ancestry of the domestic donkey, is attempting to piece together the complicated story of their heritage.[4]

A wild ass (top) and an onager (bottom), from an early 13th-century bestiary.

The other branch of wild asses, the Asiatic, covered a vast area, from the Red Sea to northern India and Tibet, and they had to adapt to different climatic conditions, from the lowland deserts of central Asia to the mountainous regions of Tibet. The main species of Asiatic asses were kiangs from Mongolia and Tibet, onagers from Syria and Persia, kulan from Turkistan and khur from India. Although these Asiatic asses have no domestic descendants, captured animals were bred with

THE ONAGER OF INDIA (*Equus hemionus*, var. *E. onager*).
(From photograph taken in India.)

The other branch of wild asses originated in Asia, such as the onager in Syria. The Asiatic asses have no domestic descendants.

domestic donkeys and with horses to produce mules. Over the years, however, the wild asses of Asia have been killed for sport and their skins, and their habitats have been eroded, so that there are very few remaining in the wild. However, their blood line remains because there was interbreeding between them and the teams of domesticated donkeys that were later to travel on the 6,000-mile 'Silk Road' across Asia to Mediterranean ports.

In the dry, hot and seemingly inhospitable and infertile rocky domain of their desert home, donkeys thrive. The donkey's body is adapted to arid environments through temperature control and water metabolism. The first strategy for heat tolerance, 'thermolability', results in reduced heat gain from the environment and, therefore, the decreased need for food and water. The normal body temperature of donkeys is 36.5 degrees Celsius but it can fluctuate from 35–39 degrees as the environmental temperature varies. This 'heat-sink', which allows the body temperature to rise above normal, is found to some extent in all animals but is most prominent in camels and donkeys. A secondary method by which the donkey's tolerance to aridity is achieved is by the economic use of water in reduced sweating and water excretion, the ability to withstand high degrees of dehydration and its remarkable capacity fully to re-hydrate quickly. Donkeys have a lower water requirement per unit of weight than any other domesticated animal except camels, and in arid areas water conservation can be vital. Donkeys can last for two to three days without water and then re-hydrate quickly without any resulting physiological damage. In extreme conditions, donkeys may withstand up to 20–25 per cent weight loss due to dehydration and then recover almost immediately from this as soon as water is available. These physiological adaptations to hot and dry conditions are what have rendered donkeys so useful to humans in similarly harsh environments.[5]

The latest genetic research suggests that the two subspecies of African wild ass were first domesticated by ancient cattle herders over 6,000 years ago, as a response to the fact that the region became more arid and rainfall increasingly unpredictable. Albano Beja-Pereira, a molecular biologist from the Centre for Investigation of Biodiversity and Genetic Resources CIBIO at the University of Porto, Portugal, sampled donkey DNA from 52 countries and found that present-day donkeys are descended from two lineages of domesticated asses in north-east Africa about 5,000 years ago, making the donkey the only significant domesticated species that originated in Africa.[6] Archaeologists and anthropologists have been interested to find out where and when donkeys were domesticated and first used by people, since it marks an important point in human history: a cultural shift from a sedentary lifestyle to a more mobile society based on trade. It was use of donkeys in particular that enabled humans to extend their worlds, to travel and to trade with different cultures.[7] Archaeologists believed that donkeys were domesticated in response to a need for transport, long after the domestication of other animals that were used in settled agriculture. Albano Beja-Pereira's research, however, proves that the domestication of the donkey occurred earlier, certainly before the domestication of the horse and the camel.

The exact context for domestication is not yet clear: there is little direct evidence about the length of the process or the timing of donkey domestication. Archaeologists had made assumptions from bone fragments unearthed from various archaeological sites; but in the initial stages of domestication, donkeys most resemble their wild ancestors so distinguishing between wild asses and domestic donkeys is difficult. Furthermore, Egyptian nobility hunted African wild ass long after donkeys were domesticated, so both occur on Dynastic Egyptian sites. Because many domestic ungulates are smaller than their wild ancestors, the earliest remains identified as donkeys were established on the basis of size and archaeological context. Archaeologists believe that the earliest domestic donkey remains date from the late fifth millennium and the first half of the fourth millennium BC in the Egyptian prehistoric settlements of El-Omari, Maadi and Hierakonpolis.[8]

The discovery in 2004 of ten complete donkey skeletons in three brick tombs adjacent to the mortuary complex of one of the founder-dynasty Egyptian kings (*c*. 3000 BC) at Abydos provide the earliest direct evidence of the use of donkeys for transport rather than for meat. These 5,000-year-old bones, which reveal extensive wear on the joints, show that the animals lived their lives transporting heavy loads. As archaeologist Fiona Marshall notes: 'This is the very dawn of the Egyptian state, the engine of which was the donkey.'[9] Marshall hopes that these donkey skeletons, dating from around 3,000 BC, will help to explain the physical and social context for the domestication of donkeys and early selection and management practices. As the use of donkeys became more widespread, their journeys across Africa were linked to the long-distance caravans that crossed the region. From there they travelled along the great trading routes to other regions of the world. As they did so, as Baker makes clear, their physiology changed – and so did human perceptions of them.

Domestic donkeys are the result of years of selective breeding and their histories therefore differ according to context. Over the centuries of association and exploitation, donkeys were interbred or cross-bred by their human masters, depending on the specific characteristics required and the different environments in which they were used. They have therefore evolved into many different breeds as they adapted to the climatic and

Donkeys have been a part of the author's life since childhood. Her two Australian donkeys, Benny and Bonny, live with her in New South Wales.

nutritional environments as they moved further north. Specific breeding goals, available fodder, climatic conditions and geographical isolation led to great diversity. The fine-limbed animal of the desert became stockier, heavier and shaggier, more able, for instance, to cope with the peat bogs of Ireland. Donkeys have been domesticated for so long that their sizes and colours vary considerably; some in India are no bigger than a Great Dane, whereas in parts of Europe they are bred with care and can attain an impressively large size.[10] The largest breeds of European

Once renowned for the breeding of large and powerful donkeys, Spain is now in danger of losing several of their breeds, for instance, these giant donkeys of Andalusia.

donkeys are found in the drier countries such as Spain, Cyprus and Malta, where 'improved' breeds of donkeys were developed by humans for specific tasks. However, many of these donkey breeds went into serious decline in the developed world with the advent of mechanization.

On the Iberian Peninsula, Catalonia and Andalusia each bred a large type of donkey that made Spain a leader in the donkey and mule breeding industry during the late eighteenth and nineteenth centuries, when their jacks were especially valued in many parts of the world. These larger breeds became the foundation of the Mammoth jacks developed in the United States when mule breeding became important in agriculture and to the economy (the first imported by George Washington in 1784). In the second half of the twentieth century the donkey population in Spain collapsed by more than a million, to the estimated 73,000 that exist today, due to the intense mechanization of agriculture that took place in the 1960s and '70s. The Andalusian is an especially large and hardy breed, thought to have descended from the giant Egyptian Pharaoh donkey, now extinct. There are probably as few as 120 Andalusian donkeys remaining, saved by enthusiasts such as 'assinologist' Pascuel Rovira Garcia. Working to retain Spanish breeds of donkeys in his refuge in Andalusia, he founded ADEBO (Associación para la Defensa del Borrico) in 1989.[11]

Other Mediterranean breeds are also dying out: for example, donkeys from Sardinia and Sicily are endangered while those from Corfu are on the verge of extinction. In Greece, between 1995 and 2005, the donkey population dropped by 96 per cent. Many of those remaining died in the devastating fires of 2007 that swept the country's south Peloponnese (where about 40 per cent of the country's donkey population live). Researchers believe that within the next ten years the population will fall below 1,000 animals.[12] Donkeys are no longer needed for agriculture because motorized vehicles have taken over, except on the island of Hydra where donkeys and mules are the only form of transport. Donkeys and mules carry everything that is needed throughout the island from bottled water to washing machines. There are approximately 1,200 donkeys and mules on the island,

A mule train outside the Old High School on the island of Hydra, led by Ilias Mastroyiannis, the island's saddle maker.

Extremely powerful mules were bred in the Poitou area of France. These magnificent animals were considered to be the best working animals of all and could command high prices.

equivalent to nearly 10 per cent of the country's total population. A conference is now held in Greece biannually, to celebrate the donkey culture of the Mediterranean and to try to reverse the disappearance of donkeys from Greece, not just as a harking back to an agrarian past but also to ensure an environmentally sustainable future for the animal.

In the 1980s I. L. Mason collected a list of 114 donkey breeds worldwide.[13] Over half of these are European, nine different breeds listed for France alone. Many of them no longer exist. However, efforts have been increased to identify, characterize and gain official recognition for the variety of breeds and in the current DAD-IS (Domestic Animal Diversity Information System) database from the Food and Agriculture Organisation of the United Nations, 185 donkey breeds are recorded. In Europe, North America and the Middle East the number of breeds was found to be much higher than in the less-developed continents where the breeding of donkeys is considered of little importance compared to their work capacity. In Europe, however, despite the efforts to identify remaining breeds, donkeys are in danger of extinction.[14]

The Poitou donkey, known in France as the Baudet de Poitou, is one of the most endangered breeds of donkey in the world today. The strongly built Poitou, with its distinctive long and shaggy coat, was highly valued, and mule breeding became an important and lucrative industry. Some historians believe that the Romans introduced the Poitou and the practice of mule breeding to the Poitou region of France. The Poitou donkey has certainly existed in France since 1016 and the breed was obviously well established in 1717 when an adviser of King Louis XV described the donkeys of Poitou as being as 'tall as large mules with joints as large and powerful as a carriage horse, covered with hair half a foot long'. The mules were used for heavy draught, as carthorses are, and were considered by many to be the finest working draught animal of all. A good Poitou mule was an extremely valuable animal and they were bred in large numbers, as many as 30,000 mules a year. They were used extensively in the First World War and were bred until the years following the Second World War. However, to feed the starving French after the devastation of two wars, hundreds of Poitou donkeys were sent to slaughterhouses. It was only in 1977 that an inventory revealed that that they were virtually extinct, with only 44 donkeys recorded. A few enthusiasts saved the breed and there are now pockets of pure-bred Poitou donkeys surviving in various parts of Europe and the United States.

The enthusiasm of hobby breeders has led to the safeguarding of a variety of breeds. The British Donkey Breed Society, founded in 1967 by Robin Borwick, aims to encourage the use of donkeys and to promote their welfare generally but it also standardizes breeding regulations. Inbreeding and lack of care

With their distinctive long and shaggy coats, these Poitou donkeys enjoy their days in a sunny spring meadow in France. Poitou donkeys were saved from extinction and are now bred in a few pockets in Europe.

in the selection of breeding stock has left the British donkey with many conformation faults. Set up in 1969 to encourage selective breeding, the Stud Book records the parentage of donkeys in order to minimize these faults.[15] The American Donkey and Mule Society also aims to keep the different breeds distinct and well bred with a strong conformation. The categories are divided into miniature, standard and mammoth donkeys: the miniatures are 90 cm (36 in.) and under, while the mammoths are more than 140 cm (56 in.) high, some as powerful and heavy as a carthorse. The miniature donkey as a breed is gaining in popularity and their numbers are growing. Their small stature and friendly nature make these little donkeys ideal companion animals. Although native to the Mediterranean islands of Sicily and Sardinia, where they are now virtually extinct, they have been extensively bred in the United States to produce the American version known as the Miniature Mediterranean Donkey. Along with a growing interest in donkeys as companion animals in some developed countries, there is now also an expanding knowledge base about donkeys from studying their behaviour in the wild.

Contemporary studies of donkeys living in the wild have shown that they form several types of social groupings. These can be family groups or 'harems' controlled by a dominant jack with several breeding jennies and their young.[16] Other structures are more fluid, with small groups coming together for a few days and then moving on. However, complex and lasting relationships form within these groups, the bond between a jenny and her foal being the strongest. Jennies are known to be the most devoted of mothers: they will go through fire to save their young. Each adult donkey has a home range and these often overlap, depending on the availability of food and water. In dry areas, donkeys occur in low numbers and range over large areas, spending 50 per cent of their day browsing dry scrub. They move at a slow walk and spend plenty of time at rest, especially during the hottest times of the day. Usually one donkey will remain on guard. In the event of danger, donkeys

This beautiful miniature donkey foal, 'Elms Lilac Blossom', was bred in Ohio; the miniature donkey is popular in the US.

Donkey head
in profile.

invariably stand and face their foe; they do not flee as horses do. They will stop, assess a situation and decide what to do: they can gallop away should the situation require that response. However, for defence, donkeys have the ability to kick with both back and front hooves. Where there are young, the adults will form a circle to protect the foals; facing inwards, they are ready to lash out with their hooves at any approaching predator.[17]

In *L'Ame du cheval,* Quenon maintains that it is the ears of a donkey that are the mirrors of his soul. As donkeys possess the largest of all equine ears, it could be argued that they possess the greatest sensibility and emotional depth. Donkeys communicate with their long, sensitive ears: for example, ears laid back signal a threat. Every instinctive reaction is reflected in the position of the ears; they are indicators of the emotions of a donkey. However, their long ears are also indicators of their physical adaptation to desert conditions, as is their excellent hearing important for desert dwellers often living long distances apart; they also act as efficient cooling systems. Both social interaction and communication are vital to donkeys since they are herd animals, and mutual grooming is an important way of establishing close contact between pairs of donkeys. They communicate with each other mainly through posture and easily identify visual clues because they have excellent sight, panoramic vision and good night vision. Because they have eyes on the sides of their head, they have a broad field of vision, including any danger that might be approaching from behind. They also communicate through sounds, such as braying, snorting, grunting and 'whuffling'. The raucous bray of donkeys enables them to keep in contact because the call carries over 3 km (2 miles). In the wild, jacks bray to signal their territories, to advertise status and to keep a group together.[18]

Still the most useful transport in rocky terrain, this donkey patiently awaits his rider in the impressive rockscapes of Petra, Jordan.

The hardy physiological characteristics and social natures of donkeys have been largely responsible for rendering them so useful to humans in many cultures over the centuries. Although no longer an integral part of life in the developed world, there is a resurgence of interest in donkeys as companion animals in different contexts, such as recreational riding and driving, carrying and packing and for farm duties. Donkeys are increasingly employed as sentinels and as shepherds to guard flocks of sheep and goats, especially in the United States, Canada and Australia. Australian graziers Bruce and Angela McLeish, for instance, turned to guard donkeys after losing 400 sheep worth $110,000 (£66,897) to wild dogs in 2007 on their Darling Downs station. Donkeys are the cheapest and longest-living animal used to guard flocks and their natural dislike of canines will mean that they defend their territory against them.[19] Donkeys are also used by farmers and pastoralists to halter break and tame young cattle and as companions for young or nervous horses. They are also increasingly being used successfully with children with varying special needs, both physical and psychological.[20] They bond especially well with the young and make good first mounts because of their steady temperaments and gaits.

Of the estimated 41 million donkeys working in the world today, however, by far the most are used for the same types of work that they have been doing since the early days of their domestication. Their most common role is for transport, whether riding, pack transport or pulling carts. They may also be used for farm work, ploughing and harvesting, and turning waterwheels and millstones. However, where they remain an invaluable aid to human survival in developing countries such as India and Mexico, their neglect and bad treatment are a cause for concern.[21] The poverty and lack of education of those who

A farmer in a poor, rural community in Jamaica takes his yam, sugar canes and bananas to market some 15 km away. Without his ancient cart and precious donkey, he would find it difficult to survive.

work with donkeys can result in their abusive treatment. The brick kilns of Pakistan, for example, may be the worst conditions in which donkeys are worked today.[22]

The milk and meat of donkeys have, at different times and places, also been valued. Since the time of their first domestication, donkey's milk has been prized for its nutritional value, and its properties are recognized today in times of increasing human allergic diseases. Biochemical analysis of donkey milk shows that it is the nearest to human milk of any other mammal. It is high in unsaturated fatty acids and vitamins A, B1, B2, B6, C, D, E and F. It was used in ancient Egypt for both its medicinal qualities and cosmetic properties. Cleopatra bathed in donkeys' milk to keep her skin youthful and radiant, and kept a stable of 300 jennies for the purpose. The Empress Poppaea, Nero's wife, kept 500 jennies so that the milk for her beauty bath never ran out. In his *Historia naturalis* Pliny the Elder (AD 23–79) reported:

> It is generally believed that asses' milk effaces wrinkles in the face, renders the skin more delicate, and preserves its whiteness: and it is a well-known fact, that some women are in the habit of washing their face with it seven hundred times daily, strictly observing that number. Poppaea, the wife of the Emperor Nero, was the first to practice this; indeed, she had sitting-baths, prepared solely with asses' milk, for which purpose whole troops of she-asses used to attend her on her journeys.

Dealing with remedies derived from animals, Pliny suggested donkey milk to combat poisonings, fever, fatigue, eye strains, weakened teeth, face wrinkles, ulcerations, asthma and certain gynaecological troubles. In seventeenth-century France, a stable of jennies was kept in the Hospice des Enfant-Assistés in Paris for the production of milk for babies whose mothers could not nurse them themselves.[23] Babies with contagious diseases who suckled straight from the jennies generally recovered. By the nineteenth century, France had many donkey-milk dairies providing the precious milk to the upper classes to drink, for beautiful skin and for various remedies. The milk has been used as a medicine against whooping cough and other respiratory tract infections and is still being employed in such ways today.[24] Small businesses are building up in Europe, once again selling donkey-milk products for cosmetic purposes.[25]

Donkey meat has also been eaten in various cultures. In both ancient Persia and Rome, for example, donkeys, along with other domestic animals, were roasted whole during feasts. William Makepeace Thackeray explained why the meat was enjoyed by the Romans: 'As an article of food, ass's flesh, as might be expected from its cleanly habits and wholesome, though at times coarse diet, is excellent eating.' He further claimed that

A mule is the offspring of a mare and a donkey jack. Known for their stamina, this pair of mules is harnessed up ready for work on a farm in the USA.

in more recent times the Bologna sausage owed its 'unsurpassable excellence to the fact that the chief ingredient . . . is from the gentle and docile ass, cleanly in his habits, cleanly in his diet and destitute of all gluttonous propensities whatever'.[26] Certainly, the market for donkey meat is still lucrative in many parts of Europe and China today. Much of the demand is met by Mexico, but there have been recent studies on the viability of producing donkey meat in Botswana, where there is an interest in a share of the market.[27] Currently, plans are under way in Australia to process meat from feral donkeys for the Chinese market.[28]

A mule and a hinny, the offspring of a horse and a donkey, from the Comte de Buffon's *Histoire naturelle* (1800).

The story of the donkey would not be complete without inclusion of the mule, the donkey's offspring – and, some would argue, the donkey's greatest contribution to human history.[29]

> Throughout history the donkey has fulfilled the role of pack animal for the trader and farmer in the Mediterranean regions but perhaps most importantly the donkey has been sire to the mule. Mules became an essential means of transport in the ancient world and remained so until the building of the railways.[30]

The Sumerians and their neighbours in western Asia were probably the first people to breed hybrid equids at the beginning of the third millennium BC. Certainly, the Hittites were breeding mules in the middle of the second millennium BC and the Bible has many references to mule breeding. Although breeding occurs between wild equids, the vast majority of mule breeding takes place under conditions of domestication. Horses were spreading from the north and donkeys from the south and they were kept together and bred together for the first time.

A mule is produced by crossing a male donkey (jack) to a female horse (mare); breeding a male horse (stallion) to a female donkey (jenny) produces a hinny. Both mules and hinnies combine the looks and virtues of their parents. Generally, a mule is more like a donkey about the head, with the body of a horse, whereas a hinny has a head more like a horse with the body of a donkey. Because the offspring of either cross is sterile, a line of horses and donkeys must be kept for breeding purposes. The reason for the sterility is that mules and hinnies have an uneven number of chromosomes – 63 – a mixture of the horse's 64 and the donkey's 62. This early use of genetic engineering is interesting now that the ethics of such human intervention is questioned, especially when the purpose of the breeding is so obviously for human benefit. There is no question of improving breeding stock; each mule or hinny is the end of that line. In fact, mules and hinnies are sexually active and both are generally castrated to render them more tractable for human handling.[31]

The mule, because it is more robust and stronger than a hinny, has been the preferred cross since Roman times and has therefore been more extensively bred. Mules are stronger than donkeys and more resistant to disease than horses. In comparison to his dam, the horse, the mule is not only less liable to sickness, he has less need to be shod, is able to survive on less nutritious feed, can travel greater distances and lives much longer; mules can work for 30 or even 40 years. Many types of mule can be bred by careful selection of breeding stock, depending on the purpose for which they are intended, whether for riding, heavy draught work or as a pack animal. Although hardier than a horse, the mule is usually not as fast and has therefore not been associated with the sport of kings, racing.[32]

Donkeys and mules (and horses) have all had important roles in human societies. Their differing physical characteristics have resulted in them being used and valued differently in different contexts and cultures: all have been bred and developed by humans for their own purposes. However, their innate temperaments have not always been understood: old adages can sometimes hold more than a grain of truth. It is held, for example, that a horse will forgive and forget, a donkey will forgive but not forget, while a mule will neither forgive nor forget. Another states that you can tell a horse what to do, ask a donkey and negotiate with a mule. How the differing temperaments and physical attributes of donkeys and mules have played out in various contexts is the subject of the following three chapters.

2 Donkeys in Human History, Mythology and Religion

Animals are fluid, divine, symbolic, and real in ancient Mediterranean traditions . . . subsequently throughout the history of Christianity, the ass is a complex and extremely significant animal.[1]

Ancient drawings and texts suggest that donkeys have had religious significance, symbolic and spiritual meaning for humans since the start of their domestication. Donkeys feature significantly in the iconography of the mythologies and religions of different cultures, especially those that originated in the Middle East, the cradle of Western culture. However, these associations are of a discreet and contradictory nature, changing as belief systems developed. Once revered and valued, donkeys were later debased and looked down upon, put in their 'proper' place, far below and in the service of humans. This chapter follows the donkeys on some of their journeys as they travelled from Africa to other countries and continents, and looks at their uses and their changing fortunes. It also concentrates on their roles, both real and symbolic, in ancient Mediterranean cultures and their subsequent impact on the religious traditions of Judaism and Christianity.

Donkeys were instrumental in the development of ancient cultures in the Middle East, especially those that grew up around the Fertile Crescent in Mesopotamia and the Levant during the Bronze Age. As people moved from being nomadic hunters and herders into settled societies based on agriculture, so the use and perception of donkeys changed. They carried the burdens

Donkeys depicted on an ancient Egyptian frieze being driven into the fields to be loaded with sacks of corn.

Donkeys carrying baskets of grain to a granary, Egypt, c. 2000 BC.

in all aspects of life in these ancient civilizations: domestic, agricultural, military and mercantile. Wild asses were a part of ancient Egyptians' lives from at least 6000 BC, before the Dynastic periods: pictures drawn on cave walls by the people of Wadi Abu Wasil show them quite clearly being hunted. Although exact dating and contexts are not yet known, other

Stone Age rock art from before 3000 BC depicts donkeys carrying goods on their backs and being driven by humans.[2] Donkeys in herds of up to 1,000 animals were originally kept as dairy animals and for their meat but became the draught and pack animals that carried the economy. In the fertile valley of the Nile, the civilization of the Old Kingdom had developed based on agriculture, in which the donkey was the only working animal available. Although used for many tasks, donkeys were especially important to the farmers who used them for ploughing, for carrying the crops, for turning the water-wheels and for transporting water.

A copper statue of a chariot being pulled by four donkeys from around 2700 BC, discovered at Tell Agrab in Iraq, shows the earliest form of the wheel, one of the greatest inventions of the Sumerian civilization: believed to be the first to have harnessed the donkey to cart and chariot. The remains of donkeys and pieces of harness have been excavated from Sumerian

Donkeys were important to the economy of ancient Sumeria, used in agriculture and trade and to pull war chariots.

burial sites in ancient Mesopotamia from around 4500 BC. Historians believe that it was the harsh environment that inspired the Sumerians' achievement, an environment in which donkeys thrived. Ancient texts make evident that donkeys were an essential part of Sumerian culture and economy, used in agriculture and trade and in the construction of cities. Although donkeys pulled chariots, they were also the common agricultural draught animals. Onagers were kept for crossbreeding with donkeys for the production of large and strong hybrids that were used to pull war chariots.

One of the ways in which donkeys were exploited by ancient civilizations was by merchants on the earliest trading routes that linked cities. Because of their hardiness in the desert conditions, donkeys were ideal for this important economic aspect. Initial interactions involved distribution of food across ancient Egypt and trade with other cultures in Africa and western Asia. Beja-Pereira notes that donkeys joined human society at about the time the first complex city-states were formed. He comments:

> I don't think it's wildly speculative to suggest that the use of donkeys, which were the first tamed transport animal, played an important role in the unification of distant cities. It marks the boundary between human societies concerned with survival and agriculture and stabilized people who wanted to explore and trade.[3]

The earliest long-distance merchant trading paths formed around 4000 BC involved the transportation of the semi-precious stone lapis lazuli from the Chagai mountains in western Pakistan to early urban settlements situated some 2,000 kilometres away in lowland Mesopotamia. Donkeys were also used in the mines in the search for precious stones and metals, as

Painted earthenware donkey, tomb goods from Northern Wei dynasty China, AD 386–534.

well as for transporting gems and spices back to Egypt. For 2,000 years the pharaohs sent donkeys to the copper mines in the Sinai to carry ore to the smelter, and to transport gold from Nubian mines.

Donkeys were sometimes employed in unusual ways on trading routes on land and, more surprisingly, at sea. The Greek historian Herodotus noted that Armenian ships sailed down the river to Babylon in boats carrying barrels of wine; in the middle of each boat stood a donkey. After the wine had been sold, the merchants dismantled the boats and loaded them onto the donkeys' backs for the return trip by land. Donkeys were also employed to enable horses to carry precious goods along trade routes. Water to keep the horses alive was carried by teams of donkeys in skin bottles slung under their stomachs in order to save evaporation in the direct sunlight.

Damascus, because of its important position on several major trade routes, including two main routes of Palestine, the 'Way of the Sea' and the 'King's Highway', became known as the 'City of Asses'. Since it was the meeting point of many trade routes and the main means of transport was the donkey, it was

The Chinese poet Tu Fu on a mule, from a hanging scroll by Kano Shigenobu (*fl. c.* 1630).

also a popular place for traders to exchange and trade donkeys. They travelled in great caravans, one reportedly comprising 3,000 donkeys, carrying all the goods in the largely desert areas stretching from the Nile to the Euphrates. Trade between Libya and Timbuktu, for example, was operating as early as 1200 BC. It was a natural meeting point for nearby West African populations with Berber, Arab and Jewish traders throughout North Africa, and thereby indirectly with traders from Europe. Right up until the nineteenth century, donkeys, along with mules and horses, were favoured on the Tabriz to Trabzon trade route, over which 45,000 animals commonly travelled three times a year. Through trading and travelling, the Egyptian donkey spread from the Nile to other areas along the many trade routes developed between western Asian, Mediterranean, Chinese and Indian societies.

Donkeys came to Greece during the rule of King Solomon (971–931 BC) having travelled the trade routes via Syria and Palestine. As they had been in Ancient Egypt, here they were also used in all areas of economic production; they carried oil, wine and grain to be loaded onto the ships; on construction sites they hauled rocks from quarries and logs from forests; they carried the goods from town to town as merchants traded their wares. They also carried people on their backs or in small carts. In agriculture they were used to thresh grain, to grind the wheat into flour and to plough the land. They were especially useful on the narrow tracks between vines on the steeply terraces hillsides in viticulture. Indeed, writers in the ancient world recognized the importance of the donkey in agricultural work of all kinds: Roman commentators later explain the donkey's particular strengths. Palladius comments on their tolerance of hard work, their sturdy natures and the fact that they need so little maintenance. Columella adds that they break down more

A donkey features large in this album leaf by Bin Xie, 1661.

Detail from a 16th-century tapestry depicting work in the vineyards; here donkeys carry dorsers loaded with grapes.

slowly than any other animal used for ploughing and that they are rarely stricken by disease. As historian Mark Griffith points out: 'Without them there would have been no food for the table or fuel for the fire; nor would the workshops, markets, and retail stores have been able to conduct their business.'[4]

The importance of donkeys to viticulture is the most likely reason that they arrived in Europe, as the Greeks brought their vines and their donkeys to their colonies all along the coast of the Mediterranean before the second millennium BC. Donkeys thrived in Spain and Italy, and the breeding of large donkeys and powerful mules was to become an important trade in those countries. Donkeys were dispersed to other countries as the Roman Empire spread. Roman armies had first to cross the Alpine passes, which they could achieve only because they had hundreds of donkeys working in trains; droves of pack donkeys were then used to supply the expanding empire. They were used in Roman colonies in agriculture, along with mules, and in the new vineyards that the Romans planted as far north as France and Germany. And from there they were taken to Britain as part of the Roman invasion.

The exploitation of donkeys in England serves to highlight their changing fortunes within one particular culture. Although they had been in Britain since the time of the Roman invasion, donkeys were more common in England from the time of the Norman Conquest and several appear in the Bayeux Tapestry. However, according to the thirteenth-century scholar Bartholomew Anglicus in *Mediaeval Lore,* donkeys were poorly perceived and harshly treated:

> The ass is fair of shape and of disposition while he is young and tender, or he pass into age. For the elder the ass is, the fouler he waxeth from day to day, and hairy and rough, and is a melancholy beast, that is cold and dry, and is therefore kindly heavy and slow, and unlusty, dull and witless and forgetful . . . For he is put to travail over-night, and is beaten with staves, and sticked and pricked with pricks, and his mouth is wrung with a bernacle, and is led hither and thither, and withdrawn from leas and pasture that is in his way oft by the refraining of the bernacle, and dieth at last after vain travails, and hath no reward after his death for the service and travail that he had living.

For 1,500 years donkeys were not greatly used or valued in agricultural work in England, in comparison to the more powerful horse. However, they were again exploited at times of war. By the sixteenth century horses had become scarce because the army had taken them as mounts. Farmers needed a replacement and donkeys were once again found to be useful in agriculture. Out in the fields in rural England, donkeys were ploughing, turning water-wheels, transporting the harvest and carrying goods to market. However, by the eighteenth century they could be found more commonly working in the expanding industrial towns, where their abuse and neglect were evident on the busy streets.

Harsh conditions continued and as late as the nineteenth century similarities can be drawn between attitudes towards donkeys and their status and treatment in Britain and in ancient Greece, especially their association with the lower classes. As Harriet Ritvo found in her investigations into British attitudes towards animals in Victorian Britain, interaction with animals reflected traditional and deeply held convictions in which animals were divided into categories that reflected the deep class divisions within that society.[5] Before the days of mechanization, the donkey was the beast of burden for those who could not afford a horse. The horse and pony were what Robin Borwick calls 'the prime movers' and were valued for their speed and working capacities. Successful tradesmen had smart ponies and traps; the unsuccessful relied on a slow, plodding, unkempt donkey.

Donkeys were used around the mines in northern England, often delivering coal to houses in the neighbourhood, and carrying the miners to and from their work. Otherwise they appear to have been the servants of those employed in the lowliest trades, hauling little carts to collect laundry, rubbish and horse manure or carrying chimney sweeps and their brushes. Donkeys were well known as milk carriers in some areas, and in London, on Saturday mornings, 2,000 donkey-barrows reportedly visited Covent Garden, driven by costermongers who sold their fruit and vegetables at the market.[6] The nineteenth-century social researcher, Henry Mayhew, reported that there were often 200 donkeys sold on a good market day. The men who stood behind the still and silent donkeys would beat them and shout at them to make them prance. However, Mayhew also maintained that most costermongers treated their 'mokes' with kindness and took pride in their appearance. They were encouraged to display their donkeys at shows and there were a variety of prizes available in different categories. Donkeys were bedecked in ribbons and their harnesses glittered with gaudily coloured ornaments.[7] This kind of treatment of donkeys was far from universal in London, however, even among costermongers.

Donkeys were often the method of transport for the poor in Victorian Britain. Artist unknown, *Seeing the Doctor*, watercolour sketch.

Perhaps the first and only time that a donkey has actually appeared in court was in London in 1822, as evidence in the case for the prosecution. This was the first legislation in the world to outlaw cruelty to animals, passed by Parliament that year. Martin's Act, as it became known because it was introduced by Richard Martin MP, a keen campaigner for animal rights, was pioneering legislation intended to 'prevent the cruel and improper use' of cattle, sheep and equines. The case aroused considerable ridicule in the press, from the public and amongst Members of Parliament. However, Martin was determined to highlight the ubiquitous cruelty to domestic animals, and the first prosecution was of Bill Burns, a costermonger, charged with cruelty to his donkey. Martin is said to have insisted that the donkey appear in court so that his wounds could be seen, and Burns became the first person to be convicted and fined for animal cruelty. A satirical print of this event, which also features the words of a song commemorating the court case, illustrates how seriously the matter was taken.[8]

Bill's Donkey then was brought into court
Who caused of course a deal of sport.
He cocked his ears and op'd his jaws
As tho' he meant to plead his cause.

Cruelty to donkeys was not a cause for concern for most people at the time. Indeed, an article in *Blackwood's Edinburgh Magazine* in 1840 reported that the donkey was 'mistreated in England, unfed, homeless, vagrant, unpitied, untended, kicked, lashed, spurred, tormented, troubled, thumped, and thrashed'.

It is nevertheless worth noting that change can happen when there is the will and motivation. A law passed in Britain in 2005 requires donkeys taking rides on Blackpool beach to work a maximum seven-hour day and to have an hour's lunch break. One British reporter considered that:

> The Brahmin-like reverence now accorded to British donkeys is a compensation for inherited guilt: we are atoning for the sins of forefathers who thrashed and starved their charges to an early grave, pausing only to conjure up more spiteful metaphors based on stubbornness, stupidity and indolence.[9]

The British nation had been regarded as one of the cruellest and least sentimental towards animals in Europe up to the early part of the nineteenth century; by the end, it was a leader in animal welfare.[10] As noted in 2003 by Tim Dowling in the *Guardian* newspaper, donkeys receive a disproportionately high percentage of charitable support from the population of the

Donkey rides along the beach are a feature of several popular coastal resorts in Britain; these colourfully decked-out four await customers.

Donkeys roam free in the New Forest in Hampshire.

UK – and raise funds with ease. The Donkey Sanctuary in Devon, which cares for 75 per cent of donkeys in the UK and runs sanctuaries in many developing countries, is now the largest organization of its kind in the world and is one of the highest earning charities in Britain, having raised £18 million in 2006. In the twenty-first century, life for donkeys in Britain has changed dramatically. It is an outstanding example of how quickly attitudes can change in a different philosophical climate.

In this overview of the donkey in human societies, it is important to consider not only how donkeys were used but also how it was that they lost respect and value and came to be ridiculed and reviled. It was largely the denigration of donkeys by the ancient Greeks and Romans that most affected their subsequent treatment in the West. This had much to do with the uses to which donkey were put, as compared to the noble horse; just as there were different classes of humans, so there were different classes of animals. Both the Greeks and the Romans depended on slave labour and donkeys came to be associated with slaves. Aristotle stated that it was natural for free men to dominate slaves and animals because both were much less rational than free men and actually benefitted from domination. Furthermore, in *Politics*, he stated that 'the use made of slaves and domesticated animals is not very different, because both, with their bodies, serve our essential needs'. Donkeys' contribution to these societies was invaluable, yet they were generally unappreciated because of their relegation to inferior status. However, their loss of status also had a great deal to do with how donkeys were represented symbolically.

Sawrey Gilpin (1733–1807), a horse and donkey in profile, aquatint.

The mule is of particular interest here as he occupies the middle ground, being half-horse and half-donkey. In some cultures, the mule falls into the lowly, slavish category of the donkey while elsewhere he aspires to the aristocratic nobility of the horse. Mules have a long and proud history stretching back thousands of years. Paintings and other depictions of mules dating back to 1400 BC decorate the walls of Egyptian tombs. Alexander the Great's funeral wagon was hauled from Babylon to Alexandria by 64 magnificent mules. This long train was arranged in sets of four, all selected for their great size and strength, each splendidly caparisoned with collars and harnesses mounted with gold and enriched with precious stones. Mule breeding was an important and sophisticated profession in antiquity, developing into a huge industry in Roman times. In his accounts of Roman agriculture, *De re rustica,* Columella placed mule breeding second only in importance to the breeding of racehorses. They were used for all of the tasks in which both horses and donkeys were employed; but, they also pulled ceremonial chariots and carriages, the preferred means of transport for the nobility.[11] Mules appear to have held a conflicting middle ground in the ancient world, mainly depending on the uses to which they were put, some considered on a level with the horse, and appreciated and well treated by their owners, others classed with the lowly donkey and treated accordingly.

In Europe in the Middle Ages, however, unlike donkeys, well-bred mules were highly regarded by the upper classes. Cardinal Wolsey (*c.* 1473–1530), richly robed in red velvet, rode on a magnificent white mule caparisoned in gold. His association with his mule was renowned, as Shakespeare describes the demise of Cardinal Wolsey in *Henry VIII* thus: 'He fell sick suddenly, and grew so ill / He could not sit his mule.' Mules were popular mounts among upper-crust ecclesiastics: tradition

In Renaissance Europe, well-bred mules were highly valued by men of wealth and leaders of the Church. In his magnificent fresco of the *Journey of the Magi* (1460), Benozzo Gozzoli depicts leaders of the Medici dynasty accompanying the Three Kings, riding on fine mules.

dictated that churchmen of the rank of bishop and above rode mules. This was a custom dating back to biblical times (when Solomon was seated on King David's mule). When the heir apparent was designated, it became common practice for them to be seated on a mule: fine, large mules symbolized dignity and honour. Even the popes, while based in Avignon during the fourteenth century, rode on mules. To breed mules of this quality, suitable for papal use, it was necessary to mate a high-class mare, such as a Lipizzaner, to a donkey jack, something that would now be thought of as heresy. A beautifully detailed

Antonio Pisanello, sketch of a caparisoned mule, *c.* 1430s. Highly prized at this time, mules were often decorated with finely made harnesses, sometimes encrusted with precious jewels and gold.

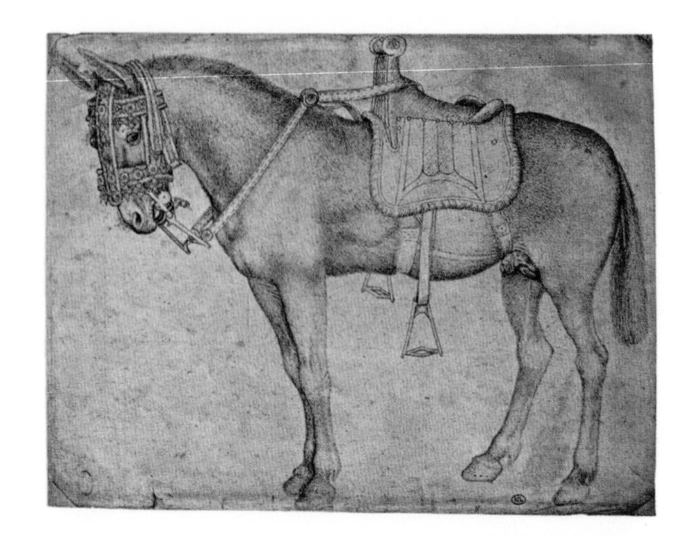

example of such a well-bred mule is depicted in *Mule* from the Vallardi album, *c.* 1430s, by Antonio Pisanello (*c.* 1395–1455). However, after the Reformation the demand for such mules ceased over most of Europe and the breeding of fine mules no longer took place. With the horse taking ascendancy, and no doubt because of having a lowly donkey for a father, the stature and status of the mule dropped accordingly.

In the early days of the Old Kingdom, animals were all-pervasive in the lives of the ancient Egyptians. The sacred and the secular were linked in Egyptian mythology and the religious significance of animals played a critical and complex role. Animals were, in general, well treated and respected and were not considered inferior to humans. Animal attributes were admired or feared and linked to the divine. The donkey had originally been so holy an animal that its ears, represented as two feathers sprouting from the end of a sceptre, became the mark of sovereignty. Representations of the ears became the prototypes of divine and royal power: they formed the power of the glyph, tokens of remembrance that all power derives from the mighty desert god, Seth. This was a good example of the physical and symbolic donkey in harmony in human perceptions, when so often throughout their history the two have been at odds.

Ten complete donkey skeletons discovered at the burial site of Abydos in Egypt point to the high regard in which donkeys were held in the days of the Old Kingdom.[12] The site of Abydos, situated in the Nile Valley 480 km (298 miles) south of Cairo, is famous as the burial place of the earliest Egyptian kings and as the cult place of the god Osiris, himself a mythic king of Egypt and ruler of the Land of the Dead. In the Old Kingdom, the focus of religion was on protecting the pharaoh, who was considered to be a descendant of Osiris, with whom he would be

A donkey among tributes being given to the Achaemenids, on a bas-relief in the Apadana Palace, Persepolis, Iran, 486–465 BC.

A maenad and a mule, from a 6th-century BC Greek vase.

reunited in the afterlife. The donkeys' burials and their location in the high-status area of the North Cemetery indicate that they were highly valued, their contributions to the daily lives of the ancient Egyptians recognized as they took their place alongside the kings of Egypt in their burial chambers. As a source of spirituality, sustenance and companionship and an essential element of the cosmic order, they were to be companions into the afterlife.[13]

The donkey was the totem for some desert tribes: ancestors of the Israelites, for example, took the donkey as their totem, in admiration for an animal that was so well adapted to survive desert conditions.[14] Tacitus, the Roman historian, records that Semitic tribes worshipped donkeys because without them they would not have been able to survive in the desert: the Hebrews used them to carry water, just as their descendants, the nomadic desert tribes, do today. The Hebrews also herded many thousands of donkeys, more than they could possibly have used, which further points to their economic value. In the nineteenth-century Thackeray noted that 'The Midianites, in their war with Israel, lost sixty-one thousand asses, and the Idumean patriarch counted one thousand she-asses as part of his wealth'.[15] Among the early nomadic tribes were the ancestors of the Israelites who did not worship Seth or sacrifice his animals, as was the Canaanites' practice. An exception to the requirement of sacrificing the first-born domestic animal was made for the donkey; a lamb was killed instead. Ways of perceiving donkeys, deeply rooted in religious traditions and practices, resulted in both positive and negative outcomes in the value placed upon them and the treatment they received.

The high regard in which donkeys had been held in the Old Kingdom in Egypt was not to continue. By the end of the Middle Kingdom the invasion by the Hyksos had caused disruption and uncertainty in Egyptian society which led, among other things, to the use of the Hebrews and their donkeys as slaves by the Egyptians. Thus began the association of donkeys with the slave class. Their symbolic association with Seth further resulted in a loss of status as new myths were wound around Seth. He was often depicted with a donkey's head, his sexual potency symbolized by the donkey, especially the erect tail and ears. Although a once powerful and revered god of the desert, Seth fell from grace. He was transformed into a dark power, the god of storms, chaos and evil. According to legend, the goddess Isis hated the donkey above all beasts because he was sacred to Seth, her persecutor and the murderer of her husband Osiris. The Greeks later linked Seth to the evil Typhon and both came to be associated with demonic forces. Plutarch recorded an Egyptian festival in which donkeys and men with Typhonic colouring, sandy like a wild ass's coat, were pushed over cliffs in retribution for Osiris's murder. To the early Egyptian Christian Copts, the donkey symbolized all that was bad, and one was chosen to bear all the sins of a community and was destroyed, usually by pushing it over a steep cliff.

The demonization of the donkey, which began in the Late Kingdom of ancient Egypt, took flight in ancient Greek and

Roman mythology. There was a time when men lived happily on earth before Zeus introduced a deadly race of women who, according to myth, were all created from ten groups of animals, all except one of which had negative characteristics. The seventh-century poet Semonides, in his *Race of Women*, categorized women according to these animal types: the donkey-woman was lazy, awkward and obstinate, with an insatiable appetite for food and sex. The donkey as sexual symbol was first illustrated in the Minoan art of Crete. It presumably derives from Seth and the Egyptians, as Egyptian influences in Minoan culture are common. Some of the central myths of Egyptian religion focused on the penis and the sexual potency of Seth was often represented through the donkey. They also appear in Minoan art-holding jugs, illustrating their association with grapes and wine, an important task in the physical world. The Greeks borrowed from Minoan culture and Middle East historian, Richard Bulliet, argues that the donkey-headed demons found in Minoan art

This ancient Greek drinking-cup shows an aroused satyr mounted on an also-aroused donkey, suggesting sexual stimulation caused by wine.

Silenus, part of the drunken retinue of Dionysus, was usually depicted as riding on a donkey or mule, as in this 5th-century BC terracotta figurine.

were the forerunners of the *silenoi* of Greek culture: half-donkey, half-human beings usually depicted with a huge phallus. These were replaced by a single minor god called Silenus, part of the retinue of the god of wine Dionysus, regularly represented as a fat, drunk old man with erect penis riding on a donkey, presumably because donkeys were essential in the cultivation of the vine.

The cult of Priapus, the Greek god of fertility, born with a massively enlarged penis, was similar to that of Dionysus, and Priapus was often depicted as a donkey. Despite these darker symbolic connotations from ancient Egyptian, Greek and Roman mythologies of evil, drunkenness and sexuality, the donkey continued to have positive symbolism for the Hebrews and, later, Christians. The donkey connects Christianity to Roman traditions. Early Christianity was influenced by Egyptian religious traditions, which had also impacted for centuries on Greek and Roman religious traditions, as well as those of the people of Israel. A melding of these Egyptian, Canaanite, Jewish, Christian

A Greek terra-cotta figurine of a woman on a mule, 1st century BC.

and Islamic traditions meant that the 'sacred aura of the donkey far exceeded that of any other domestic animal in the region'.[16]

Early practices of donkey veneration by the Hebrews may help to explain why the enemies of the Jews in Roman times were able to describe them as donkey-worshippers. Democritus claimed that the Jews worshipped the head of a golden ass, to which they supposedly sacrificed a fattened Greek every seven years. The calumny of onolatry, or ass-worship, attributed by Tacitus and other writers to the Jews, was later transferred to Christians who shared much of the Jewish tradition. Tertullian recorded that one day a man appeared on the streets of Carthage carrying a figure with hoofs and the ears of a donkey clothed in a toga and bearing the sign 'The god of the Christians, conceived by an ass'. Although a Christian himself, he appeared to find this mockery amusing rather than offensive. Many early writings strongly suggest that the worship of the God of the early Christians was associated with the head and/or penis of the donkey. Christian and Jewish scholars have studied these stories and found that they provide 'extensive evidence of donkey veneration in the ancient world and numerous clues to the peculiar association of the ass with Judaism and Christianity'.[17]

The three great monotheistic religions born in the Middle East, Judaism, Christianity and Islam, developed over many centuries and from a wide variety of cultures and traditions, contain many conflicting and contradictory voices with regard to the value of animals. However, although all recognize that animals are creatures of God, they all relegate them to low status in comparison with humans. Scriptures from all three

A Greek bronze mule-head ornament for a couch, 1st century BC.

have rules for the treatment of domestic animals such as donkeys – but these have been variously interpreted over time, largely on the understanding that their purpose was to serve humans. The donkey was certainly an important element in all three religions, its role symbolic as well as physical. Donkeys carried humans and their burdens in everyday life in this world, but they were also believed to have special qualities that connected them to the spiritual world.

Some biblical stories attest to the fact that donkeys were believed to have prophetic abilities. Perhaps the most famous of these references is the biblical story of Balaam's ass (Numbers 22: 21–35), the only animal actually to speak in the Bible (apart from the serpent). The donkey saw the angel of the Lord while the human did not and God chose to speak through a donkey to show Balaam the error of his ways. According to Jewish tradition, Balaam's ass was special and she was hidden away and cared for until such time as she would be called on to carry

A Greek drinking vessel in the shape of a donkey's head, 5th century BC.

Rembrandt van Rijn, *Balaam's Ass*, c. 1620s, oil on canvas. God chose to speak through a donkey to show Balaam the error of his ways; she is the only one to see and hear the Angel of the Lord.

the long-awaited Messiah. Islamic traditions contain numerous examples of talking donkeys that speak to their masters in human language. Muhammed is said to have talked with his personal donkey, Yofir, a descendant of Noah's donkey. According to Muslim legend, only a favoured few animals are allowed in the Garden of Allah, two of which are Balaam's ass and Muhammed's donkey.

Donkeys figure prominently in the scriptures of all three religions. One concordance to the Bible, for example, lists 153 references to the donkey, more than any other animal.[18] They are often mentioned in the Bible as being ridden by persons of wealth: judges and their sons rode on asses while kings and

A Greek terra-cotta figurine of a mule carrying baskets, 4th century BC.

their sons rode on mules. Indeed, most Jews rode upon donkeys, while those of the royal line of the house of King David rode upon mules, indicating their high status. Solomon, for instance, rode upon a mule when he was proclaimed king. The links between donkeys and mules and prophets are also strong in Islamic literature. When Muhammed died, the last of God's messengers according to Islamic beliefs, Yofir was said to have run head-first into a well to commit suicide to end the hereditary line of prophets' donkeys stretching back to Noah. She did not want to be ridden by any other than the Prophet. Muhammed's mule, Duldul, did not commit suicide, however, and was passed on to his cousin and son-in-law, who was the founder of what later became known as the Shi'ite branch of Islam, in which Duldul became a revered animal. Shi'ite tradition drew on the stories of Muhammed and his donkey to incorporate into their ideology. Shi'ite Imams chose to ride donkeys to emulate the prophets as evidence of their humility.[19]

Donkeys became exemplars of piety and humility in Christianity because of their associations with Jesus and their ability to recognize the revelation of the Incarnation. The donkey was also the bearer of Christ. The centrality of the donkey is evident in these iconic images. Two donkeys are mentioned symbolically in the Gospel story – one coming from the north and bearing Mary to Bethlehem, where, according to legend, an ass and ox stood over the crib; the other taking her to Egypt to escape the slaughter of the innocents. Jesus' triumphant entry into Jerusalem on the back of a donkey is one of the most enduring images of Western culture and it is usually described as a symbol of his humility. As has been explained, there was a tradition of the Messiah riding upon the lowly ass. Kings and messiahs did ride on donkeys, at least those coming in peace; conquering emperors rode on fine chargers. The ties with Christianity are so strong that legend has it that the cross on the donkey's back came from the shadow of the Crucifixion, a living symbol that the donkey has carried through the centuries. It could be argued that Christianity once again raised the status of

Senufo masks, such as this mule–head mask from the Ivory Coast region, combine the features of animals and humans and are believed to have the power to aid communication between the living and their dead ancestors.

Nativity depicted in a stained-glass window in a church in Lamezia, southern Italy.

the donkey: when others failed to recognize Jesus, the donkey was there, helping, supporting and worshipping.

The bray of the donkey has been written about over the centuries and has been a major cause for the scorn heaped upon the animal. The raucous and discordant sound has encouraged ridicule and confirmed donkeys' clownish and inferior status. Phosphor Mallam considered that it was the bray, so easily mocked, that set people against donkeys.

> When that long drawn rocketing peal, in swelling volume of nasal treble and guttural base, rings and echoes . . . and the tragic-comic clangour is traced to the distorted throat and bared gums of an incongruous moke, risibility is tickled and man gives way to irresistible laughter.[20]

The bray has entered the folklore and fables of many countries and is believed to be an omen, both for good and evil. It has been considered the voice of the devil and a fool, and of an angel and a nightingale. It has even played a part in history. Herodotus describes a time when the brays of the donkeys of the Persians turned the battle in their favour, frightening the horses of the Scythians and causing chaos in their ranks.[21] According to Persian folklore, donkeys bray because they see the devil. It was therefore considered an unpleasant and strident sound, one resembling the voice of an evil spirit. Philosopher Theodore Lessing, however, thought the donkey represented the uncomplaining mass of working people and that the bray expressed the miseries and misfortunes of humanity.[22]

3 Donkeys and Mules Colonize the Americas, Australia and South Africa

> Two thirds of the New World would hardly have been civilized yet without him . . . the burro has been the corner stone of history and the father of civilization . . . he has developed more mines than all the railroads in the world; and has been to innumerable millions of pioneers the whole engine of success. Yet in these dwindling days it has become the fashion to sneer at him.[1]

Here we concentrate on examples of the contribution of donkeys to modern Western history, when Europeans used donkeys and mules in the colonization of the Americas and Australia. South Africa is also included in this story of colonization, a particular event forcefully illustrating the association of the donkey with the poor and marginalized within a society. The contributions of donkeys to colonial societies have all too often been sidelined or overlooked in official records and histories. This is particularly the case in Australia, where it is difficult to find records of donkeys in colonial times, and where their descendants are now persecuted as 'vermin' by government authorities. The results of humans transporting animals to lands where they were not indigenous has become a cause for concern in some post-colonial societies, where donkeys are now playing out their newly ascribed role as 'out of place' in these environments.

The donkeys' journeys alongside humans had taken them from the Middle East to Europe via several routes. When the Moors arrived in Spain in 711, they shaped the history and culture of the Iberian Peninsula, making many changes to the way of life for the people. Mules were used instead of oxen in agriculture, and donkeys were used to transport produce to market. In 1496 Queen Isabella of Spain gathered her forces to oust the Moors. A total of 60,000 mules and donkeys, perhaps one of the largest pack trains ever assembled for war in Europe, carried supplies for the expedition. When Isabella looked westwards for gold and spices, Christopher Columbus was sent in search of the Spice Islands. Instead, he discovered America; and when the conquistadors explored and settled the New World they took donkeys, along with horses and mules. The first arrived with Columbus in 1495, although a number had been taken to the West Indies a few years previously. Historians claim that the transportation of these animals across the sea was one of the greatest contributions that the Spanish made to the West.[2] Records of shipments of donkeys to the New World are few but on board Columbus's ship on his second voyage there were four jacks and two jennies. Early arrivals of donkeys came to both North and South America, to Mexico and Brazil. North-eastern Brazil was invaded early and the donkey became, and remained, its most valuable domestic animal. The donkey's hardiness and resistance to the climatic conditions enabled humans to settle and develop the region. As Frank Brookshier explained, although donkeys accompanied the conquistadors to the New World, they came to be more closely identified with those who settled there, the miners, farmers and traders.[3]

Donkeys also arrived early in Venezuela, most likely shipped there from the island of Margarita, where Spanish colonizers had established cattle ranches and coastal fisheries; donkeys carried salt for both industries. Surprisingly, the mountains of Peru also became home to the donkey, once they had adapted to the cold climate of the high Andes, where they still work to this day. Llamas, donkeys and mules, as well as human slaves, all worked at the rich mine at Cerro de Pasco, carrying loads of silver ore down the mountainside. European explorers and colonizers took donkeys with them to virtually every region of South America. Darwin commented during his travels in 1835 on the reliance on donkeys for domestic tasks by the people of Chile. Cattle ranchers of Argentina used donkeys, along with horses and mules, and all remain indispensable to the gauchos working on the pampas today. In other parts of South America, donkeys continue to carry crops and produce such as potatoes, corn, wheat, vegetables, bread, peanuts, cassava, cotton and tobacco to market, as well as transporting essential firewood, water and people.

Fifty years after the arrival of Columbus, the donkey was well established in Mexico and had become an important aspect of the economy by the middle of the sixteenth century. Among the first animals to travel eastward over what was to become the Santa Fe Trail were donkeys and mules. This important transportation route running through central North America connected Missouri and New Mexico. The earliest traders on the trail used donkeys and mules to carry Spanish blankets. Donkeys were one of the few animals that could survive the arid conditions of Santa Fe. They were used mainly to collect firewood from the surrounding hills. Mules, who quickly proved their superiority in terms of strength, endurance and pulling power, became an institution on the trail, along which many hardships and deprivations, including lack of water, were experienced. However, the contradictory and capricious attitudes of humans towards donkeys were witnessed at Santa Fe, as in other parts of the world. Once they were no longer needed, due to increasing mechanization, the donkeys were turned loose to fend for themselves in the arid environment.[4]

A Zuni Indian of North America collects a load of fuel on his donkey.

The first Spanish colonizers of the United States of America took mules and donkeys with them to the mines of the Rio Grande, and in the south-west donkeys came to be most closely associated with prospectors and miners. The first arrived with the expedition led by the Spanish colonizer Juan de Onate in 1598. Inventories taken of the expedition recorded numerous livestock, including mules and donkeys. The various colonists had their own stock; one inventory, for example, included 30 mules and donkeys. As early as 1609, pack trains of donkeys were travelling between Chihuahua and the Rio Grande bringing supplies and equipment and usually returning with salt. As they pushed further west, donkeys quickly became the prospector's friend, carrying all their equipment and employing their quiet and friendly natures to become companions. Many did not require leading ropes but would simply follow their masters. Their hardiness and surefootedness made them especially valuable in rocky and mountainous regions.

One of the best-known donkeys was 'Prunes' of Colorado, who had worked all the mines around Fairplay. He had toiled underground for so long that when he was finally retired he was unable to eat grass, so he visited people in their homes and was fed scraps. He was especially loved by the children of the town and followed them to school. When he died a monument was built for him, and later Robert Sherwood, who had worked with Prunes for 40 years, requested he be buried beside the monument, so that donkey and miner would rest together forever. Another of Colorado's famous mining donkeys was the faithful companion of Winfield Scott, 'patient, plodding, and silent as he'. They searched in vain for gold for seventeen years until they finally discovered one of the richest gold mines ever found.[5] Donkeys played their part in these strikes and in mining development as a whole, working in the tunnels, transporting equipment, carrying baskets of rocks and ore. Some donkeys

Donkeys carried pioneers' tools as they searched for gold. They became the prospectors' friends and companions and followed their masters without the need of leading ropes.

Donkeys were an important aspect of the mining industry, especially in the Rocky Mountains. These donkeys await their loads outside a silver mine in Colorado.

received recognition for their contribution; others lost their lives. But most were appreciated by the prospectors.

There is a wealth of folklore associated with burros in the American West, where pioneers and prospectors told yarns about the feats performed by their animal companions. Burros reportedly saved their master's lives on many occasions by finding help when they lay injured, water when they were dying of thirst or by finding their way home when they were lost in the desert. In 1896 Charles Lummis (1859–1928) quotes one prospector, George Harvey, and the wonderful stories he told of his six burros. Harvey obviously loved and admired them and received faithful service in return. He believed that although donkeys were

> Lacking in the blue blooded lineage of the horse . . . unlike the horse, they have brains. A horse may be loyal and obedient but a burro is loyal and obedient and smart. He exacts only one condition from his owner; he wants

him to show him what he wants done. Then he will do his best to do it.[6]

Another prospector asserted in 1925 that:

> The burro is the only living animal that can stand the hardship of the American prospector's life. A horse or mule would die of privation and despair if he were to suffer half the hardships, grief and toil that the burro thrives under. If it were not for the burro, the American Desert today would still be an unreclaimed and unexplored waste. Thunder and lightning, intense heat or bitter cold, the sirocco sand storms and white dust of the alkali waste are all met with patience and courage by the burro.[7]

The Spanish mule also proved valuable to the early settlers and became an important part of their culture. Several histories maintain that the mule played a more significant role in the development of the West than did the horse.[8] Perhaps most famous are the twenty mule teams and the great wagons that carried thousands of tons of Borax from Death Valley to the

Donkeys also carried the timber posts to act as props in the mines.

William Henry Jackson, a surveyor on a mule in Colorado, *c.* 1870s, photograph.

railroad 165 miles away.[9] This hot, dry and desolate desert region could not be worked by any other animal (camels were not used in the United States). The hardy and strong mules, the offspring of mustang mares and Spanish jacks, also became popular with many cowboys for working stock. The mules were the main choice of riding animal for most of the pioneers of the West, the cavalry troopers as well as the cowboys, despite the fact that films invariably depict these iconic characters on horseback, presumably to give them higher status. Mules were also to become the backbone of the farming industry with large teams pulling great ploughs across the open plains, cultivating ground and harvesting crops.

The great success of the mule in the United States was largely due to the foresight, example and skill of George Washington, the country's first president (1789–97). A keen farmer and owner of a large plantation, Mount Vernon in Virginia, Washington was always searching to improve agricultural procedures and products. He knew that progress for America depended on strong and hardy work animals, and was convinced that this meant the mule. The problem was obtaining proper breeding stock. The best donkey jacks in the world resided in Spain. Of remarkable strength and proportions, the monarchy had long guarded these animals by prohibiting their export. Nevertheless, by 1784, Charles III of Spain had arranged for two fine Andalusian jacks to be dispatched to the American president. One died during the voyage but the other, named Royal Gift by Washington, went on to sire many hundreds of high-quality mules, which as Washington had planned did indeed help to build America. Throughout his life, Washington bred mules and he was responsible for breeding what is today known as American Mammoth Jackstock.

The mule was used in farming, commerce, transportation, logging, mining and riding. Mules also towed great barges laden with farm produce along the Ohio canal system to markets in the south and east as settlers headed for new lands. Mules were in great demand and were imported from Catalonia, Malta, Andalusia, Italy and France. A main reason for the extensive mule industry was the expansion of the cotton-growing industry.[10] Breeding centres developed to provide work mules for the cotton fields of the Old South, while mule teams pulled the reaping machines over the vast fields of wheat on the prairies.[11] Mule breeding increased dramatically between 1870 and 1900 and 'nothing to compare with it in the world's history has been seen'.[12] In 1889, for instance, 117,000 mule foals were born and 330,000 mules were sold from the main breeding states of Missouri, Tennessee and Kentucky for agricultural use in other states.[13]

Although there was a serious decline in numbers when the tractor replaced the mule teams in agriculture, mules generally retain their popularity in the United States. There is renewed interest in mules for recreation and competition and at least one enthusiast believes that 'While the mule contributes less than one percent of the nation's work power today (as opposed to 79 per cent in 1850), many dealers think the old mule breeders will soon return to business.'[14] Mules have also become the mascot of the US Army: carefully bred and beautifully cared-for mules are paraded at public occasions.[15] Enthusiast and author William Long believes that the mule is

Mules were still used in the mining industry well into the 20th century. This miner and his mule worked at the American Radiator Mine in Pennsylvania.

Once bred and utilized in tremendous numbers before the advent of mechanization in America, mules are once again gaining in popularity. This pair is getting ready for a ploughing exhibition at Louisville, Kentucky.

> the most successful hybrid that man ever developed. Bred one generation at a time for more than 3,000 years his patient labour was essential to the development of the mechanized world that has made him obsolete.[16]

Although the mule's qualities were not universally recognized, as can be seen in the reputed political insult that 'The Democratic Party is like a mule – without pride of ancestry or hope of posterity', there is little doubt that the mule was an outstanding favourite.

This interest, use and appreciation did not spread to Australia. Certainly, Britain was well known for its lack of interest in mules and prejudice against them as an inferior hybrid of the horse. This viewpoint is clearly expressed, for instance, in Arthur Vernon's *The History and Romance of the Horse*:

> There is no dignity about a mule. There is no charm, no glamour, no spirit, no gallantry. There is something tragic

> about a mule. It is not a horse, nor is it an ass. Lower than a mongrel, it is just a hybrid, an energized hash, a fore-doomed slave and a biological travesty.[17]

This sort of attitude could account for some of the prejudices that many of the British colonists brought with them to Australia, with the result that mules, although appreciated by some, never really took hold as they had in the United States. In contrast, admiration of the horse for its majestic nobility was illustrated in the flood of paintings and prints of that animal. The wealthy kept thoroughbreds and enjoyed the aristocratic pursuit of horse racing, but even 'those who followed the plough preferred horses to other draft animals, no matter how strong or cheap to maintain'.[18]

Draught animals were indispensable to the British colonizers of Australia since they were responsible for hauling people and their goods during the appropriation of land in the search for pasture and minerals. Without them, the pioneering journeys could not have been made, or the land settled. Donkeys first arrived in very small numbers into New South Wales in 1793; but their importation into South Australia in 1866 was the start of their real use to the early colonists. The power of draught animals was vital to the European transformation of Australia to a 'modern country' and as a major supplier of wheat, wool, meat and minerals to the international economy. This happened in a relatively short period of time, starting with the arrival from England of the First Fleet with its human and non-human animal cargo in 1788. By the middle of the twentieth century, after teams of draught animals had hauled their loads across the vast continent for over a hundred years, the transformation had been achieved.

Donkeys and mules were to prove especially useful in the semi-arid regions because of their ability to survive in dry conditions

Donkeys were the 'unsung heroes' of colonial Australia: large teams hauled goods across the Outback.This team are transporting wool from Cordillo Downs to Innamincka in South Australia.

and to negotiate hard and rocky ground, unsuitable for all other draught animals. They adapted easily to the types and amount of feed available and were able to travel for tremendously long distances. They were not given commercial fodder and could survive on what their handlers described as 'nothing'. Donkeys can survive on the dry scrub that is to be found in these arid, semi-desert areas. Long periods of drought saw donkeys and mules not only surviving but also working on the stations, while all other stock was removed. Another important factor in the use of donkeys and mules was their resistance to disease. They were unaffected by conditions that either killed or rendered other draught animals unable to work. It was a while, however, before these characteristics were understood and appreciated, mainly by the teamsters who drove them or stock workers who rode them. In records and histories their efforts remain largely overlooked. Donkeys were the unsung heroes of the colonization of the harsh Australian outback.[19]

The first few donkeys arriving in the more fertile colony of New South Wales in 1793 came aboard the *Shah Hormuzear,* which sailed from India. Several more arrived over the next few years but what became of them is unclear, as are the reasons why more were not imported at that time. It would seem that donkeys did not have an important role in the establishment or economy of the colony. The Australian Agricultural Company imported the first mules into New South Wales in 1840, from Chile. Transporting wool and materials by bullock dray was a slow and laborious business, especially in wet and boggy ground. The commissioner of the company, Phillip Parker King, decided to introduce mules because they could transport goods along a route too rocky for bullocks and horses. The Chilean mules proved their superiority over both horses and bullocks on this terrain. They were able to carry 3 tons of flour over the route in five weeks, whereas bullock drays had taken five months, while their advantage over horses was their ability to negotiate stony ground without becoming lame, and to live off scrub along the way.[20]

Donkeys were shipped from Afghanistan in 1866 to work in the expanding pastoral industry of South Australia. The donkeys were imported with camels in response to the need for draught animals that required less fodder than horses or cattle, as the European occupation of the dry, central regions of the continent began. The first team of donkeys was operating by 1869 along the rough and stony tracks between Port Augusta and Lake Hope, with sparse vegetation and little water. Mules were originally imported there from Chile in 1853 to work in the mining and smelting industries on the Burra copper fields, although they later also worked on pastoral properties. Donkeys and mules, as camels, were exploited in the arid regions in which horses and bullocks perished; they were rarely seen in the more fertile and more populated areas of Australia.

Donkeys became important in a region where summer day-time temperatures can be well over 38 degrees Celsius for weeks on end, their efficient body cooling systems an obvious asset.[21] They were also easy to train, could be driven by anybody and were cheap to outfit. The donkey teams were used to carry goods to and from the outback settlements, the pastoral properties and mining towns. They carried all the necessities out to the isolated stations and the arrival of the donkey team was an important event in station life. They hauled great wagons loaded with wool bales back to the ports and railheads. They were also used in the building of roads, railways and telegraph routes. Donkeys were used on the pastoral stations for carting gear, transferring windmills, digging dams, carrying stone for building work and carting fodder and wood. They were therefore known to be versatile and amenable and were used for riding, mustering, as pack animals, in teams and for pulling carts and buggies.[22] Because of their heavy fodder requirements, an alternative to horses and bullocks was sought. Donkeys and mules provided much less competition for fodder. Pastoralists, concerned to run the greatest number of sheep on their stations, recognized this fact, especially in times of drought.[23]

During the devastating drought years of the early twentieth century donkeys were central to the survival of several stations, especially those owned by the Beltana Pastoral Company, the largest and most successful group of pastoral stations in the colony of South Australia. One reporter was told that the Beltana 'would never have had a sheep on them if it had not been for the mules, donkeys and camels'.[24] Stud donkeys were imported from Spain in the early 1900s, which proved an important milestone in the breeding of both donkeys and mules and in the improvement of the quality of the animals produced. Beltana's breeding programmes provided the donkeys and mules that were to form the foundation of the teams for the next few decades until the introduction of motorized transport, and long after in some of the more remote and harsh areas of the continent. Statistics show that numbers of donkeys increased rapidly: the breeding programmes were highly successful. By 1920 they were working in considerable numbers on the pastoral properties of South Australia, where recorded numbers were 3,115 donkeys and 1,063 mules.[25]

As prospectors and pioneers moved west and north from South Australia in search of gold and land, donkeys became an integral part of the mining and pastoral industries of Western Australia. During the early gold rushes in the 1880s, donkeys arrived from South Australia more slowly and in far

A donkey team being loaded with wool bales at Mannahill, South Australia, ready for their journey to the nearest railhead.

fewer numbers than camels, but by the early 1900s they were in general use throughout the goldfields and the pastoral industry. Their numbers increased dramatically as the number of camels began to decline. Donkeys proved vital in the far north and north-western districts of Western Australia when Kimberley horse disease, or 'walkabout disease', decimated the horse population and when the cattle tick destroyed thousands of cattle in the colony in the late 1800s/early 1900s. 'Walkabout disease', caused by the poisonous plant *Crotalaria retusa*, resulted in horses suffering chronic liver damage and severe nervous symptoms, including continuous aimless walking and banging into objects.[26] Donkeys were affected by neither of these disasters.

The vastness of Australia meant that teams of draught animals had to travel great distances and the stories of the donkeys and their teamsters in the outback show their considerable hardiness and tenacity. Some of these journeys took many weeks, even months, so it was important that the wagons were loaded only with the goods needed. Any food and water that had to be carried was for the people and 'pack horses' – the donkeys fended for themselves and drank at the wells where they camped. If feed was good, they could go for two days without water. The ease with which the donkeys were caught and handled was another important factor in their favour for their human handlers. Several teamsters noted that donkeys were 'good campers' and hung around the camp at night, unlike horses and camels that could often take half a day or more to round up. Once rounded up in the morning, all a teamster had to do was put his hand on the rump of one of the donkeys and it took its correct place in the team and waited to be harnessed. Then the teamster walked along the line suggesting that the donkeys pull. Unlike horses, which jib (pull up short and refuse to go on) if they cannot move the load, donkeys keep on pulling until they do, so it is not necessary for them all to start pulling at the same moment.[27]

Cliff Finn, a teamster from South Australia with a story typical of many others, recalled that, in the 1920s, he drove a team of between 15 and 27 donkeys and that the wagons they used, built in Port Augusta, were four-wheelers, the back wheels being much larger than the front; the box section was about 4 metres (13 ft) long and carried about 8,000 kilograms (17,637 lb). The teams travelled about 32 km (20 miles) a day and the round trips he took carting between the stations took between three and six months. He reported that the donkeys chosen by the teamsters to be leaders of the teams were the most intelligent and obedient but that they also possessed keenness to work and affection for their teamsters. He said that the drivers knew all their donkeys individually and that each had a name.

The best known of the donkey teamsters of Western Australia was George Kinivan, who kept the stations supplied with goods in the Kimberley region between 1915 and 1940. He reportedly had a team of 72 donkeys that pulled an enormous six-wheeled dray.

> He yoked these seventy two donkeys into a single team to drag his home made dray, loaded with twelve tons of groceries, from the port of Wyndham some 300 miles to Soakage Creek well back among the Kimberley cattle stations. The journey usually occupied six weeks which was good going in those days when there were no made roads and the track, as it does today, led over country unbelievably rough and broken, across watercourses deep in sand, and over endless stony ridges where the wheels left no impression.[28]

Kinivan maintained that despite the increasing use of motor vehicles and the advent of aircraft, donkeys did not disappear on the Kimberley tracks for many years. He explained that donkeys could cover terrain that would defeat both horses and motors: 'they fill a place which could not be taken by any other animal or quite as efficiently by motors'.[29] However, the era of the large donkey teams was drawing to a close and it was a hard time for many teamsters.

The donkey teams were no longer economically viable to cart wool to the ports and railheads, but some continued working around the stations, while others were set free. The teamsters were too fond of their donkeys to consider shooting them. They had been their companions and their livelihood over the years and many were sad to see them go. These donkeys thrived in the wild. Their numbers increased rapidly and they formed into herds, which became 'pests' to the pastoralists, reputedly competing with cattle and sheep for food and water. The hardy characteristics of donkeys that rendered them so valuable to European colonizers are the same qualities that lead to their successful survival in the wild. Yet they are now socially constructed as 'exotic invaders' that do not belong, 'vermin' to be exterminated.

The response to and fate of 'feral' donkeys that multiplied rapidly and began to take over land differs between the United States and Australia. The gestation period for a donkey is 365 to 370 days. This ensures advanced physical coordination skills at birth so that the foal can escape from predators and run with the herd very soon after birth. The jenny begins her reproduction cycle at three years of age and, although she has only one foal a year, she can reproduce for 30 years; in ideal conditions, a donkey herd can therefore multiply substantially. In the United States, 'feral' donkeys were reported in the Grand Canyon as early as 1884.[30] By the 1930s they were a cause of concern to land

The 'Kona Nightingale' is a breed of free-roaming donkey found in the Kailua-Kona region of the Hawaiian Islands descended from a population of former worker donkeys.

resource managers in most western states. They were believed to be threatening cattle-grazing ranges, so Federal agencies and private citizens attempted to control or eradicate the wild donkeys by shooting or poisoning them. Hunters decimated the herds, but there was protest from some sections of the public. In 1952 legislation was passed making it illegal to shoot wild donkeys in Death Valley in California, where the greatest numbers roamed. A sanctuary was set up for their safety where they could live out their days in peace.

In 1971 the United States Senate and House of Representatives passed Public Law 92–195 which declared that

> wild free roaming horses and burros are living symbols of the historic and pioneer spirit of the West; that they contribute to the diversity of life forms within the Nation

> and enrich the lives of the American people . . . [and] they shall be protected from capture, branding, harassment and death.[31]

In fact, wild donkeys have been removed from a number of National Parks, most notably the Grand Canyon, and agencies for and against the donkey still battle in various states; but it would seem that those who wish to preserve the donkey as an important player in American history are winning with their 'Adopt a wild horse or burro' scheme, run by the Bureau of Land Management. Despite attempts to revoke the protections afforded by the Wild Free-Roaming Horses and Burros Act of 1971, it was reaffirmed unanimously in the House of Representatives in May 2006, with the passage of an amendment prohibiting tax payers' money being used to sell or slaughter wild horses and burros. In March 2007 the Natural Resources Committee reported on this decision. The chairman declared:

> Americans have always championed their survival, and they expect that these creatures will be protected. To allow them to be sacrificed and slaughtered represents great disrespect to the will of the American people and is an affront to our nation's history.[32]

In Australia, on the other hand, government agencies are intent on the eradication of 'feral' donkeys and the majority of the general public seems to have little knowledge of or interest in the matter. As one Australian donkey enthusiast so aptly put it:

> It seems a tragic reflection . . . that animals that were once invaluable – who played such a vital part in the opening up of this great country – should, in a generation or

two, be thought of as vermin and hunted down in their thousands to be killed for the bounty paid on their ears or as pet meat.[33]

In the Australian outback, the pastoral system and environmental factors combined to produce favourable conditions for donkeys, once domesticated, to survive in the wild. It was estimated in 2004 that there were some 300,000 wild horses and five million wild donkeys in Australia.[34] Although donkeys are not visible in most populated areas and are therefore not considered major economic 'pests' country-wide, they are considered to be a serious environmental threat in parts of the northwest, causing soil erosion and damaging vegetation with their hard hoofs. It is alleged that they cause pasture degradation through overgrazing, competing with domestic stock, and are therefore unpopular with pastoralists. However, little research has been carried out to substantiate these claims. Tremendous numbers of 'feral' donkeys have been reported in some areas and there have been concerted efforts to reduce their numbers. For example, in 1988 it was reported that large herds often outnumbered cattle on some stations in the Kimberley region, which carried 5,000 cattle and 10,000 donkeys. Over the preceding decade, 164,000 donkeys had been shot in the area.[35]

No one really knows how many hundreds of thousands have been shot on the immense stations of Western Australia and the Northern Territory: the areas are too vast and difficult to monitor. One of the biggest runs in the Northern Territory, Victoria River Downs, for example, is the size of Denmark: 41,155 sq. km (25,572 sq. miles). Donkeys in this region are now slaughtered in their thousands, shot by marksmen from helicopters. As the donkeys have retreated further from humans and cattle into rocky and inaccessible areas, their bodies are left to rot where they fall. It is too difficult and expensive to retrieve the carcasses for the pet meat trade.

The latest weapon, both simple and effective, in the eradication campaign is the Judas Collar Programme, which works on the social instinct of the donkeys. Several donkeys, usually jennies, are captured through trapping or tranquillizer darting and a radio collar fitted. The 'judas jennies' are then released to join target groups in the area, leading shooters to the herds. As one herd is shot out, the donkey moves on, leading the marksmen to the next group, and so on. Some claim that this method is responsible for reducing the population of 'feral' donkeys and horses by over half a million since the 1970s.[36] Following the introduction of the Judas Collar Programme in 1994 in southern Kimberley, the eradication of wild donkeys was deemed by government authorities to be successful. Over 270 radio collars were fitted and five years later the Agricultural Protection Board reported that they were over half way to achieving their aim of complete eradication.[37] Figures for 2007 show that 25,520 donkeys were shot on West Kimberley pastoral leases. The programme is ongoing on several of the stations in the area.[38]

The views of pastoralists, environmentalists and animal advocates differ over the question of whether it is acceptable for humans to shoot donkeys from helicopters. There are too many stakeholders involved for there to be an easy solution to this problem. There is little doubt that since 1788 European occupation, population increases, land clearing and technological advancement have led to enormous environmental degradation. Australia has become home to a wide variety of Eurasian plants and animals and a great deal of damage has been caused by introduced species. However, the greatest harm is clearly attributable to human activity.[39] As the human population increases, the more the environment is degraded. Ethically, exploitation is no longer considered an appropriate human response to the environment; but the exploitation of certain animals is ignored. If the animals are further labelled as 'pests' or 'vermin', then humans can dismiss their suffering and distress as somehow not relevant. Whether donkeys should be shot because they are no longer of use or economic benefit to humans needs to be debated in the public arena before they are exterminated. Their real 'sin' is to compete with the more economically viable cattle on the vast stations of the Northern Territory.

This situation was played out in a different context in South Africa. Although donkeys originated in Africa, they were not indigenous to South Africa and were slow to penetrate south. Donkeys arrived with European colonists, with the Dutch occupation of the Cape: the first landed in 1656, from where they spread northwards. They were later brought by the London Missionary Society as pack animals to carry the post in 1858.[40] Although there is little information available about the origins of donkeys in South Africa, there is little doubt that they arrived with all the European cultural baggage on their backs. Slowness and smallness were interpreted as inferiority, and wilfulness as stupidity: few of the donkey's positive qualities were recognized. At first, donkeys were not much used apart from asbestos mining by Whites. Perhaps because Whites had little use for donkeys, they became cheaper and started to appear on Black reserves and proved practical as transport and haulage animals for poorer members of society.

Donkeys became important during the nineteenth century in agriculture and for transport: the use of donkeys as pack animals or for pulling carts was popular in rural areas, enabling small-scale farmers to participate in the market economy. At the beginning of the twentieth century there were about one million donkeys and mules employed in the country. Although

One of the most common uses of donkeys in many poor communities, such as these in rural Ethiopia, is to fetch and carry firewood.

their use and popularity declined in mining and in large-scale agriculture, their importance remained in small-scale farming. As more influential South Africans stopped using donkeys, so their reputation diminished in government circles and information regarding their management disappeared from official documents and educational materials. There were also attempts made by the government to reduce their numbers. Here, as in most societies, donkeys have been associated with poverty and low status; and they came to be linked to lack of progress.[41] However, they played an important role in the lives of people who were marginalized by wider development policies and their ability to survive the environmental conditions ensured their continuing use. This is one of the major reasons for the resurgence of interest in donkeys in present-day rural Africa.[42]

The donkey as a victim in the history of South Africa is one aspect that is explored by Nancy Jacobs in her book *Environment, Power and Justice,* which presents a socio-environmental history of the Kuruman district of the Kalahri Thornveld. In her research for evidence of people interacting with their environment, Jacobs cites the donkey as an important player in these relationships: 'The donkey, often viewed as a comical beast of burden, articulated relations between poor people, the environment and the colonial economy.'[43]

Indigenous tribes in Kuruman acquired donkeys in the early 1900s as they were better adapted to the semi-arid and impoverished environment than cattle and they became very useful to the poor. Donkeys were cheaper than cattle, healthier and had many subsistence uses, especially for carrying the maize harvest in carts. Specialized carts were built to carry water, wood, gravel and sand for brick making: the owners of these carts could make a living by charging for their services. Although not a favourite food, donkeys could also be slaughtered and eaten, and their milk was used as a medicine for sick children.

However, according to the official construction, donkeys became an environmental menace and in the mid-twentieth century state programmes were implemented to reduce their numbers. The segregation policy of the 1960s and '70s, known as 'Separate Development', had a detrimental effect on donkeys and their owners, especially in the state of Bophuthatswana. Elite cattle-ranchers benefitted from state assistance while donkeys had little economic value and were perceived as slow and outdated. A crisis arose with the great drought of 1983 when cattle died in greater numbers than donkeys and the government blamed the donkeys for using valuable resources that should be saved for the cattle. What followed was one of the most dramatic and traumatic animal events in the recent history of Apartheid: the Great Bophuthatswana Donkey Massacre of 1983.[44]

Stock reduction was one of the most contentious aspects of 'Betterment', a government policy that involved the forced removal of people and animals according to modern principles of agriculture and conservation. It also portrayed African farmers and herders as incompetent and responsible for the destruction of the environment. Farmers in Kuruman had limited stock so there was not a problem with excess animals; the concern was with 'inferior animals'. Donkeys were considered incompatible with efficient, modern agricultural progress. The official verdict on donkeys was overwhelmingly negative; their supposed detrimental environmental impact was not researched and their utility to poor villagers overlooked; but this made little difference to the state criticisms of donkeys and the decision to cull them.[45]

The first cull in the district in 1949 involved the sale of the donkeys, mainly to the bone-meal factory. In 1953 a proclamation decreed that on all reserves in the area donkey numbers were to be limited. Villagers did not openly oppose these culls; some obviously agreed with the 'received wisdom' about donkeys, including the environmental damage they caused. The ownership of donkeys carried no prestige, unlike the ownership of cattle. Furthermore, donkeys were especially useful to women, and women played no part in village meetings where their fate was discussed. However, many people who recognized the ongoing value of donkeys to their lives hid them in their homes during culls, rather than accepting a one-off cash payment.

Donkey numbers continued to be discussed during the 1960s and '70s and limited numbers per household were agreed upon since the majority of rural people engaged in subsistence production and did not participate in commercial cattle raising. The 1980s saw devastating droughts, which threatened the commercial interests of the cattle ranchers. In 1983 a government decree announced that all 'surplus' donkeys were to be exterminated. Those people who proved that their donkeys were necessary could keep four. 'What followed grew out of the precedent of earlier donkey control, but it had astounding and unparalleled vehemence.'[46]

Members of the Bophuthatswana army and police descended on the villages and shot over 10,000 donkeys. Some donkeys had been gathered into yards ready to be counted, and were an easy target for the shooters. The violence of the shooting distressed witnesses because the donkeys were shot anywhere in the body, many times, and many suffered greatly. A few people managed to hide their donkeys but many reported losing their whole herd. One man who was out with his team of four donkeys had them shot out from under him. He had to get friends to help him to move the carcasses and to haul his cart home.

The donkey killing led to greater poverty for those affected; and it did little to save the cattle, which could not survive on the impoverished and diseased land, with or without donkeys. Despite this, by the 1990s donkeys had a poor image, especially among the wealthy or urban dwellers. They came to be regarded as pests or vermin: unproductive and harmful animals that had to be controlled, even eliminated. Wealthy cattle owners held power in the government, which decided that donkeys should be killed. However, the Bophuthatswana massacre later became associated with Apartheid and was condemned in all quarters. Therefore, ironically, it accorded value to donkeys. Today in the Kuruman area there is popular support for donkeys, lauding their moral and economic significance to the poor, and to Christianity and to democracy. All the negativity surrounding donkeys is denied, including their supposed environmental impact. There is a high Christian population in Kuruman and the donkey massacre is held to be a serious moral transgression because of the donkey's biblical significance. There is an appreciation of the significance of donkeys for the poor or lower classes and the killing is viewed as a class-based injustice.[47]

This photo of children riding on a donkey cart in Kitobo, Tanzania, shows a typical type of harness, adapted from whatever materials are available.

Two monuments that were erected in honour of the work done by donkeys also show how they came to symbolize the great divisions of class and race in South Africa.[48] White people erected them at around the same time as the massacre: one in Upington and one in Pietersberg. In Upington, donkeys powered the water pumps for commercial fruit production, while in Pietersberg they carried rocks during the gold rush of the late nineteenth century. Both monuments have inscriptions acknowledging the donkey's contribution to the economy. In both places the donkeys contributed to white human progress and grazed on land belonging to their owners.

Over one million people benefit from donkeys in the new South Africa and there is an increasing demand for them among many small-scale farmers. There is renewed appreciation of their sturdy survival qualities, especially due to the drought that has affected many southern African countries in recent years. Donkeys are usually cheaper than cattle as they are smaller and are valued for their ease of management, especially by women and children, and for their greater endurance and longevity.[49] Despite the fact that the authorities continue to claim that there are too many donkeys in parts of South Africa, in the same areas small-scale farmers report a shortage. Indeed, it is a shock to many Africans that feral donkeys are shot in other parts of the world. Today donkeys are used largely in cultivation, for weeding and ploughing, and for transport, and it seems that demand will continue, even as South Africa develops. Authorities need to take a positive position and assist in finding ways to use donkeys more efficiently; and to help donkeys themselves, for if people manage them appropriately they can continue to support poor families in rural South Africa, as they always have done.[50]

The stories of the donkeys in these three colonial settings illustrate their importance to humans in harsh conditions. They have been the helpmates to some of the poorest in society, those struggling to make a basic living in lean circumstances. Despite their obvious value to those who worked with them, those with power, including landowners and government agencies, have denigrated donkeys. They have been represented in different contexts as inferior and lacking in progress; as harmful and a threat to 'more important' animals; and as vermin to be exterminated.

4 Donkeys and Mules at War

To all the animals who suffered and perished in the Great War knowing nothing of the cause, looking forward to no final victory, filled only with love, faith and loyalty, they endured much and died for us.[1]

Arguably, the greatest sacrifices that donkeys and mules have made for their human masters have been in times of war. They have been involved in human warfare since the time of their domestication. The same qualities of strength, endurance and cheap upkeep that made them valuable in peacetime likewise made them a military asset. Early Bronze Age Syrians first employed donkeys to pull war-carts to carry supplies, while ancient Roman and Greek armies exploited both donkeys and mules for pack work and riding and for pulling chariots. They have remained an important means of transport in most theatres of war ever since, especially in inaccessible, difficult and rocky terrain, and in harsh climates. Although those contexts and circumstances vary, whether the tropical jungles of Burma, the mud of France, the sandstorms of the Middle East or the snow of the Alps, donkeys and mules have proved critical in many engagements.

The Neolithic period in the ancient Near East saw great changes in economic and social structures, which eventually led to military warfare. Two developments that had a great impact on military history were the domestication of animals and the formation of two social systems, farming and nomadic. Conflict between the two systems was to form the basis of military history in the Near East until the twentieth century AD, beginning

A panel from the Sumerian 'Standard of Ur' shows war chariots pulled by four donkeys.

in Sumeria with the donkey and the wheel.[2] On the Sumerian artefact known as the *Standard of Ur*, dating from around 2600–2400 BC, five war-carts are depicted being drawn by donkeys and donkey/onager hybrids.[3] This early representation of a Sumerian army, possibly engaged in a border skirmish due to constant warring factions in the developing city states, illustrates the importance of donkeys to their conflicts. Two of the carts are being pulled at walking pace in a procession; the other three appear to be moving at speed, the donkeys trampling bodies of the enemy under their hoofs as they surge forward. However, these early chariots must have been difficult to manoeuvre in battle and it is likely that they were more commonly used for transport and defence, rather than attack.

From the earliest conflicts armies have used animals to carry the basic requirements of war that could not be shouldered by soldiers: transporting equipment as pack animals was the most important use of donkeys in the logistics of warfare. The loads ranged from food and water to weapons and clothing; occasionally, donkeys were also used as mounts. The military use of donkeys allowed armies to remain in combat for longer and to campaign further afield, proving especially useful for marches in desert regions. The successful organization of all the transport needed for an army, including the moving of the sick and wounded, often decided the outcome of the battle. Campaigns could fail if insufficient planning went into the feeding and care of the animals: once they died from exhaustion and starvation, the human soldiers had little chance of success.

A major military advantage of using donkeys is that they can feed on sparse pasture and fodder of the poorest quality, even leaves and thistles. For example, when Pompey's troops prevented Caesar's animals from foraging, donkeys were fed on seaweed washed in fresh water and mixed with grass.[4] In the Roman army, horses were the preferred mounts for the cavalry and officers; but the army could not have operated without mules and donkeys. The high fodder ration required for horses was a major reason why donkeys and mules were the preferred pack or draught animals. Accounts of the amount that donkeys can carry vary considerably. Roman sources suggest that the average load for a donkey was just over 100 kg (220 lb) and that they sometimes carried in excess of 175 kg (386 lb), which would seem excessive overloading when the larger mules were carrying 140 kg (308 lb).[5]

One of the greatest challenges to the Roman army was transporting water to the men and animals, especially if the campaign was being fought in desert conditions. There are several records of numerous pack animals carrying water in specially made water-skins. Pompey, for example, ordered water for his troops to be carried in 10,000 skins when crossing a desert region, while Herod is reported to have supplied water for Roman troops in their march across the Sinai that was carried on the backs of donkeys.[6] Although slow, covering 4 km (2 miles) an hour and 24 km (15 miles) in a day, donkeys are particularly efficient pack animals for their size if they are not overburdened or overworked. This was often not the case, however, and they were worked until they died; then others would be requisitioned from the local populations to take their place. The huge Roman Empire depended entirely on horses, mules, donkeys and oxen for all its land transport and obtaining sufficient donkeys and mules to carry the legionaries' arms and baggage round the empire was a continual logistical problem, which was to become a factor in its decline.[7]

Since the Roman army first used them, mules have been considered the best pack animal in warfare almost everywhere. However, it is only more recently, when animals have been taken into consideration as an important factor in combat, that their contribution has been recognized and recorded. In his article about the mules used in Burma in the Second World War, for instance, First Lieutenant Don L. Thrapp maintained:

Known for their great strength and endurance, mules have been a favourite means of transport in all theatres of war.

Edwin Forbes, 'A played-out mule in hospital' in the American Civil War, sketch, 1864.

> The average mule is one of the most intelligent, and certainly one of the most sure-footed, animals in the world. He can see a trail where a man can see nothing but rock. If left to his own devices he will never stumble, rarely slip or bog himself down, and almost never hurt himself. When, however, he is led by a man he can perhaps get into more trouble than any other creature on the face of the globe, and although his difficulty is directly attributable to his inexperienced leader, the animal gets the blame. We occasionally lost animals over the sides of mountains, in rivers or bogs, but we would have lost not a single one had they been free to choose their own way.[8]

It is in their interaction with their inexperienced handlers that mules so often come to grief, as did many of those used in the American Civil War (1861–5). Mules were sometimes used for the cavalry, the most famous example of which was in 1863 when Colonel Abel Streight led 1,500 men mounted on mules

Mules played an important role in war. They are represented in this exotic pageant, sketched by Rembrandt, held in The Hague in 1638 as part of the festivities surrounding a royal marriage.

An Australian Light horseman and his donkey in Palestine during the First World War.

into Georgia. This campaign was a failure as the Confederates defeated the army; but the mules were in poor condition, ill and half-wild, and they therefore broke down, forcing the men to walk. It was reported that they were taken into training at too young an age, received cruel and brutal treatment at the hands of incompetent and ignorant wagon masters, that they were routinely slaughtered by enemy troops, that thousands died from starvation and overwork, and that the starving soldiers often ate them. Transportation was the prime use to which mules were put in the American Civil War and, as historian Albert Castel claims, the Federal control of the most important mule-breeding states of Missouri, Kentucky and most of Tennessee was an important factor in the defeat of the South.[9] Once the armies moved away from the rivers and railheads, they relied on mules for transporting all the supplies, either pulling wagons or carrying packs. In the Potomac campaign alone 15,000 mules were used in this way.

In modern Western warfare, both mules and donkeys have been exploited in all major campaigns: they were used extensively in the Boer War (1899–1902) and in the First World War. Although mechanized road transport was available during the Boer War, animals still played an important role in the logistics of moving the army. According to military historian, Alfred Thayler Mahan, the transportation required was 'unprecedented, and its success unsurpassed in military history'.[10] Animals were drawn from across the British Empire, as well as from Europe and the Americas. A sum of 360,000 horses out of a total of 519,000 had to be shipped into South Africa, while 106,000 mules and donkeys out of a total of 151,000 were also brought into the region. The logistical needs of the war put Britain's shipping capacity under strain, so as many animals as possible were sourced locally, and 45,000 mules and donkeys and 159,000 horses were obtained from within South Africa.

It is commonly assumed that after mechanization animals were no longer needed to facilitate warfare, but this is certainly not the case.[11] Although motorized transport was beginning to take over by the First World War, the military understood that

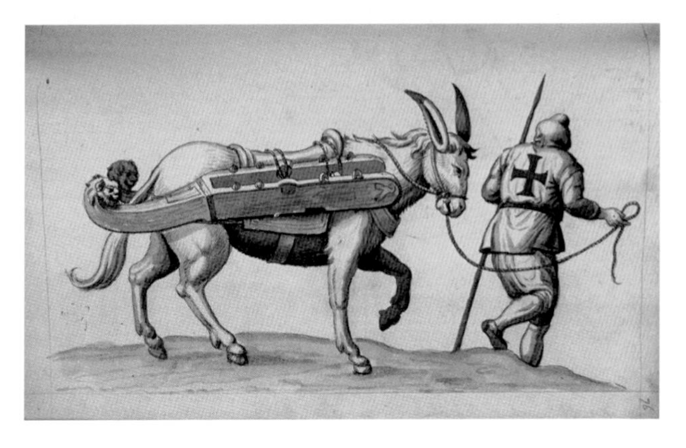

A mule carrying a cannon chassis, from a 1580s manuscript depicting military materiel.

animal transport remained the only viable alternative in many areas. The British Army, which had vast experience with and knowledge of horses but little of mules, had to turn to those units that had served in India, where mules had proved invaluable on the rugged North-West Frontier. The use of mules to carry 'screw guns' for the Mountain Artillery in the freezing and hazardous conditions became well known.[12] The gunners were proud of their mules and many stories were told about their amazing stamina and common sense.[13]

According to E. H. Baynes, the mule's foremost triumph was that achieved on the Italian front during the First World War where he was at his best on 'the steep, zigzag trails leading up to the Italian stores or gun positions':

> There is no animal who more surely repays all that is done for him. With well-cared for, well-disciplined mules, marches of between fourteen to sixteen hours duration were on many occasions accomplished over the most difficult and dangerous paths.[14]

The strongest were used to carry cannon. There were 24 mules in each battery and these were divided into groups of six, the parts of a cannon divided between them. These parts were loaded very carefully, on their backs and on either side, secured with specially designed frames and harnesses. The Italians, as the Indians, held mules in high esteem, unlike many other nations – and the mule was to become their greatest ally during the war. Mules were an essential mode of transportation in Italy due to the mountainous terrain and lack of adequate roads; and the Italians were considered to be amongst the best muleteers. The confidence that was necessary between man and mule worked particularly well as they fought together over the Alps.

An Australian soldier feeds an exhausted mule on the road at Fricourt in Picardy, during the Battle of the Somme, 1916.

The muleteers were often ex-military men who knew the area and were accustomed to the terrain. Every division of the army needed mules and there were over 200 with each battalion of Alpine troops – magnificent Poitou mules and fine imports from the United States, large Spanish mules and the smaller Sicilian mules all worked for the soldiers, who knew how to treat them.

Mules endured terrible conditions in the trenches of France on the Western Front. They delivered most of the ammunition at Passchendaele over virtually impassable ground that gradually transformed into lakes of deep mud. Many mules drowned in the mud and many hundreds drowned in shell holes. Some died from their wounds from shells, others from the effects of poison gas. However, their general good health won accolades from all quarters. Many officers were said to prefer mules to horses for all purposes because of their powers of endurance and resistance to bad conditions and their consequent length of life at the Front. Few fell sick and they were 'incredibly brave under fire', while 'their staunchness, their strength, their blind trust in their stoic drivers' was 'wonderful to behold'.[15]

Donkeys also served with all the Allied armies in France. One incident that illustrates their stoic endurance tells of a British officer who was touched and impressed by a donkey that he met on the road in a French donkey transport line. As it passed, he noticed one little donkey had no ears. On enquiry, the driver explained that an exploding shell had destroyed them and at the same time blinded the donkey, yet still he walked on with his load.[16] Although not as strong and therefore not as widely used as mules, donkeys were used as pack transport. Some carried food in great panniers to the soldiers on the front lines; they were small enough to weave their way along the trenches, allowing the distribution of rations as they went. They made friends with the soldiers, sometimes being taken into the trenches for mutual warmth. Many soldiers, including those who fought at Gallipoli in Turkey, reported on the affection they felt for the animals.

Lieutenant Colonel Carbery of the New Zealand Army Medical Corps reported that the beach at Gallipoli was 'a mass of men, mules, munitions and shrapnel; and the most deadly of these were the mules'.[17] Claims made about mules and their contributions to the war effort were contradictory; soldiers either loved them or hated them. Soldiers who did not know mules or how to treat them believed that they were vicious, obstinate and would kick at the smallest provocation. Converts knew that it was only the inexperience of handlers that caused the intelligent mule to play up. Once there was understanding and respect between them, mules were seen as unsurpassed in the tasks they could perform and there were many tributes to their bravery, stamina, hardiness and perseverance. For example, members of the Sikh Regiments at Gallipoli refused to take leave because they could not bear to entrust their mules, who worked tenaciously for them as trusted handlers, to anyone else. Accounts of the offensive contain reports of fatally wounded mules struggling on with their packs, refusing to drop dead until they had completed the task.

In 1915 there were 1,889 mules at Anzac Cove. They had little or no shelter from the bombardments and many hundreds were killed. Major William Johnstone of the Second Field Artillery Brigade described one such event in his diary in May 1915:

> The beach got a beautiful pounding this morning – the shells burst in the line of mules and blew them to rags . . . They were collecting the scattered mules and shooting the badly wounded ones . . . One mule had all the flesh off its neck and others had legs smashed and hanging on a tendon. Hundreds of mules must have been killed or wounded during that half hour's shelling.

At full strength, the Indian Mule Corps had eight troops of 96 mules each, used to provide transport of ammunition, entrenching tools and medical and signalling equipment to the front lines.[18] Transporting goods from the beaches to the soldiers was an ongoing challenge during the campaign. Those who served at Gallipoli were aware how much they owed to the mules and their Indian drivers for the supplies of guns and ammunition, food and water that they carried up razor-sharp cliffs. War correspondent Charles Bean noted: 'What would we have done without "Johnnie" and his sturdy little mules? Horse or motor transport could not have faced the difficulties of Anzac.'[19]

Donkeys, on the other hand, served a dual purpose, as transport and companions. Although they were employed as carriers, their gentle natures found them appreciated as pets and mascots by the ANZAC (Australian and New Zealand Army Corps) soldiers. Bean contrasted donkeys with mules, noting that the latter could do more work but 'the donkeys are the favourites with the men on account of their temper'.[20] They generally did the lighter work, carting water, meals or biscuits, but they also carried guns. General Birdwood, concerned about the problem of transporting water to the troops on the front lines, reported that he obtained about 100 small donkeys from the island of Imbros to carry tins of water. A New Zealand soldier reported that they brought several donkeys with them on the troop ship SS *Goslar* to test drinking water, since it was believed that donkeys would not drink poisoned water.[21]

But perhaps the role for which donkeys are best remembered at Gallipoli is as bearers of wounded soldiers. Donkeys worked with stretcher-bearers from the very start of the campaign. One of those bearers, John Simpson Kirkpatrick, was to become a legend, an icon in Australian history. Images of Simpson and the donkey, quietly going about their business of helping wounded soldiers amidst the chaos of war, frequently appear in Australian iconography. However, there were several ANZAC soldiers who used donkeys to carry wounded soldiers: Private William Henry of the New Zealand Medical Corps, for example, was probably the first to organize two donkeys for evacuating the wounded. Subsequently, other members of the unit used donkeys, especially Richard Alexander Henderson, who continued to do so long after Simpson had been shot. The donkeys were trained to carry a man up to the operating table and then back out of the tent. They were immediately recognized to be calm in a crisis and were much in demand. They may not have been as strong as the mule but, as far as the ANZAC soldiers were concerned, they were far more appealing, amenable and easy to train.

Horace Moore-Jones, *John Simpson Kirkpatrick Leading a Donkey along a Cliff Path Carrying a Wounded Soldier*, photolithograph, 1918. 'Simpson' is, in fact, Richard Alexander Henderson of the New Zealand Medical Corps.

Donkeys and mules remained indispensable in certain regions even after improved mechanization during the Second World War; and they continue to be used in particular conflict zones to the present day, despite the latest technology and sophisticated weaponry. Mules were, for example, used for keeping lines of communication open in Burma in the Second World War for the combined Commonwealth and Chinese forces. General Orde Wingate raised a guerilla army to operate behind enemy lines, preparing his brigade for two enemies, the jungle and the Japanese. This training took place in the jungles of central India and prepared the men for column and bivouac life, jungle warfare, river crossings and the care and handling of mules. The mules arrived unbroken and unhandled and had to be trained in four weeks before being flown to Burma. One soldier explained how they trained in packing the mules with their machine guns and mortars, also in leading the mules, hitching them for the night and actually moving with them on night marches:

> These were big Missouri mules and there was a small group of servicemen who apparently had been farm boys used to working with the animals. They gave us the instruction on packing and loading the packs – 90-pound heavy leather and canvas affairs with metal strips to strengthen them – and it took two of us to lift the pack onto the mule's back.[22]

The mules were airlifted in, usually accompanied by a muleteer. The American Air Command Force that was attached to the Chindits used C-47 Dakota planes to pull large gliders, with a carrying capacity of twelve soldiers and four mules, into the air. On one operation they lifted 1,500 soldiers and 100 mules behind the Japanese lines.[23] One plan involved parachuting mules in where replacements were needed immediately. An

Pack mules of the US Army Signal Corps carrying storage batteries for a field wireless telegraph.

airman recalls packing an assortment of extremely unhappy mules into a Dakota and flying with them to the front lines.

> These particular animals were unloaded at a makeshift landing strip, but other mules were rigged with parachutes and heaved out in midair. After an ill-fated first attempt, when the animals stiffed their legs and were injured on landing, subsequent drops were done with drugged mules.[24]

The mules carried heavy weapons, ammunition, radios and medical supplies over treacherous mountain trails impassable for any other transport. Indeed, Earl Mountbatten, Supreme Allied Commander, reputedly commented that it was the mule and the Dakota that won the Burma campaign.

The first operation included over a 1,000 mules. The second operation, which became known as the Mars Task Force, had less as there was more reliance on air supplies. Each column had about 56 mules providing transport: ten were required to carry the radio equipment, including batteries, generators and petrol, and the remainder carried other heavy equipment, weapons and supplies. Wingate commanded his columns with a radio mounted on a mule and he supplied them by planes that flew in from Assam.[25] All supplies came by plane but although the airdrop to the Chindits included fodder for the mules, suitable food often proved a problem and the mules had to eat bamboo and banana leaves. Because they had to work constantly without rest or relief, they were frequently exhausted in the humidity; long marches, often with saddle sores and diseases of the feet because of the wet ground, meant they suffered considerably. They were injured by shells and by poisonous vegetation, and many died of their wounds or disease.

> It was painful to a mule with a sore back just to put a saddle on him. It hurt the drivers to have to do it and then load the gun on as well. Admiration for our mules had always been enormous. Now, seeing the true extent of their steadiness and patience and their continuing endurance, it was boundless. Men developed close relationship with their mules – they could often be heard at night talking quietly to them.[26]

The march routine was usually about an hour's walk and then a ten-minute break. The mules slipped and struggled as they climbed up steep parts of the narrow trail. Soldiers found the going very difficult and appreciated what the mules achieved. Mules whose vocal chords were cut could also be used to move silently through the jungle behind enemy lines at night. Hundreds of 'stealth mules' underwent the Hobday operation, this process of de-voicing a dreaded task among ordinary soldiers who had to tie up and hold down the mules.[27]

In his memoirs as a front-line doctor in an infantry battalion fighting its way from Naples in Italy to the Brenner Pass in Austria, Klaus Huebner often refers to the mule trains. He was witness to the careful treatment that the mules received and the sorrow the men felt when they lost any of their animals. On one occasion the entire mule train bringing supplies to the soldiers was destroyed and he commented:

> The Italian mule skinners are hysterical and make no effort to collect any stray animals. They cry and shout and run off weeping in all directions. To treat them is impossible. None of them will hold still long enough to be bandaged. They scramble off the mountain, leaving a trail of blood behind them.[28]

Huebner later added that the 'Indian colonials' handled the mule trains and that they did a wonderful job, so there were obviously both Italian and Indian muleteers operating the teams. However, those 'mule trains' consisted of donkeys as well, because he noted that small African donkeys received excellent care from their Indian handlers. Once the heavy burdens had been taken from the donkeys, they were led down to the creek to be washed down, before being rubbed dry and covered with burlap blankets for the night.[29]

From the rocky Alps of Italy to the muddy and steep terrain of Papua New Guinea, mules and donkeys were taken to assist infantry battalions in whichever combat zone they were needed. One of the lesser known and perhaps least successful uses of donkeys was with Australian forces in Papua New Guinea on the Kokoda Trail. The terrain was rugged and treacherous and

A gunner of the 2/7th Field Regiment prepares to load an ammunition pack on to a donkey of the 2nd Australia Pack Transport Company in Kaira, Queensland, in 1943.

airdrops were not feasible, so all supplies had to be carried on men's backs. As the epic struggle for the Kokoda Trail reached its height, the solution of transport problems was critical to success. Since donkeys had been used as pack animals in many mountainous parts of the world, the idea of using them here seemed logical. In 1942, 400 wild donkeys were mustered from the Kimberley in Western Australia, trained, and shipped to Papua New Guinea. A second group was subsequently mustered in South Australia to be sent for training, although this was not a particularly successful endeavour. The donkeys already sent could not adapt well to the muddy conditions of Kokoda and so were not as useful as had been hoped. Therefore many of the second shipment of donkeys were not utilized, and they ended up being taken to the Abattoirs Digestor (or Boiling Down Works) where they were 'humanely destroyed' and converted into fertiliser.[30]

Trained army mules and donkeys can provide unexpected benefits for humans during peacetime. Highly trained mules of the Pakistani Army have been crucial to their operations since the time of the Royal Indian Army under the Raj in the northern mountainous regions and Kashmir. Just like the men, the mules go through rigorous selection and training procedures and are highly valued once trained. They can go where no man or machine can. If necessary, the mules can carry rations and ammunition to the forward posts on their own, once they know the way. These specially trained units of mules were instrumental in the survival of hundreds of victims of the earthquake that devastated parts of northern Pakistan in 2005. Many mules had been killed in the earthquake but those that survived were mobilized to carry supplies to inaccessible areas, sometimes without human assistance.

Many countries, including the United States and Britain, continue to use pack animals to support special operations in mountainous regions. They have, for instance, been brought back into the United States military to help fight 'America's War on Terror' and are often the chosen method of transportation for elite Special Operations forces where battles are fought in mountainous and treacherous terrain, inaccessible to tanks and other vehicles.[31] A US Army Special Operations manual provides instructions for the use of pack animals in training and combat missions and outlines the techniques of animal pack transport and for operating pack-animal units, claiming 'to recapture some of the expertise and techniques that have been lost in the United States Army over the last 50 years'. The manual describes the characteristics of mules and donkeys as well as guidelines for their health and welfare. Donkeys, handlers are told, have a strong sense of survival. If they decide something is dangerous they will not do it – they are smart enough to know what is not

Donkeys continue to be employed in the field today. Here soldiers from the US 173rd Airborne Brigade Combat Team and an Afghan National Army soldier patrol on foot while accompanied by two donkeys on their way to a village in rural Afghanistan.

manageable. Their propensity to freeze when frightened rather than run from danger is seen as a great asset – one passed on to the mule. Yet with trust and confidence in their handlers, donkeys will accomplish whatever task is required of them.[32]

The Special Operations animal support services teach soldiers how to handle the animals and pack them with specialized weaponry. Master Sergeant Larry Jones reported that the practice of shipping the animals overseas from the United States to combat areas was a mistake. The Missouri mules that were used in the first Afghan Conflict frightened the locals, because they were so large that they found them difficult to load. The US Army decided that it was better to use and train local mules and donkeys where possible. Although a devoted horseman, Sergeant Jones found mules to be the best pack animals of all and he recounted a time when a string of mules fell about 45 metres (148 ft) from a sharp cliff, landing upside down. They were unhurt but could not get up because of their heavy packs, but did not panic and continued quietly on their way once their

A heavily-laden donkey in war-torn Afghanistan.

handlers were able to free them. He also explained how the Special Operations forces rely on the mule's instincts:

> His sense of smell and hearing are so much more acute than those of the human. It's like having another patrol buddy with you on a mission. He has good senses so we just watch him.

It is interesting to note that the author of this article nevertheless writes disparagingly of 'the common old mule', and concludes with the 'good news' that the horse is once again being used by the cavalry: 'the noble horse is helping America hunt down enemies world wide, once again serving with distinction'.[33] It seems that old prejudices are still with us.

Both the Allied forces and the Taliban use donkeys in the ongoing conflict in Afghanistan. Afghanistan's terrain is one of the major challenges facing Coalition troops, where the enemy is hidden in mountain cave and tunnel complexes and there are no roads and few footpaths. Donkeys are able to survive in the arid conditions, getting urgent supplies to the soldiers in inaccessible areas. Canadian forces, for instance, are using a team of 30 specially trained donkeys to deliver crucial supplies of water and ammunition. Taliban insurgents also use donkeys as moving bombs. In one incident in 2008 a booby-trapped donkey was blown up in Kandahar, killing one policeman and wounding three others. In another, a donkey laden with 30 kilograms of explosives, and with several land-mines strapped to its back, was intercepted before it could be detonated.[34] Donkeys are also employed in the complicated logistics needed to hold elections in the war-torn country. In mountainous areas inaccessible to all vehicles, teams of donkeys weighed down by ballot boxes bigger than themselves carry the necessary materials to isolated polling stations.[35]

Any history of human warfare is also a history of the tremendous numbers of animals sacrificed in these conflicts: they have suffered death, deprivation and terror alongside humans. It is somewhat ironic that it was in such tragic circumstances that these animals have perhaps received the most appreciation and attention. It may be that humans forget their 'superiority' during such devastation, when they suffer alongside those non-human animals taken into battle with them, their lives so often depending on their valiant efforts.

5 Donkeys in Literature, Film and Art

When fishes flew and forests walk'd
And figs grew upon thorn,
Some moment when the moon was blood
Then surely was I born;

With monstrous head and sickening cry
And ears like errant wings,
The devils walking parody
On all four-footed things

The tatter'd outlaw of the earth,
Of ancient crooked will;
Starve, scourge, deride me: I am dumb,
I keep my secret still.

Fools! For I also had my hour;
One far fierce hour and sweet:
There was a shout about my ears,
And palms before my feet.[1]

Western cultural representations of donkeys serve to highlight the confused and contradictory attitudes humans have towards them, as is clearly illustrated in G. K. Chesterton's 1920 poem 'The Donkey'. The donkey is perceived as a primitive beast, an ugly misfit and a caricature of an animal from a distant primeval world: 'The devil's walking parody / On all four-footed things.' Everything about the donkey is mocked, except its capacity to remain true to itself in spite of ridicule and harsh treatment. According to Chesterton, this is based on their knowledge that they are indeed special. Jesus' triumphant entry into Jerusalem is one of the most enduring Christian symbols in Western culture.

Eadweard Muybridge's pioneering photographic work captured the movement of the mule, published in an epic portfolio of 781 folio prints, *Animal Locomotion* (1887).

The image became popular during the Middle Ages, when the donkey was associated with Christianity and was represented in a more positive light, as a symbol of patience, humility, suffering and service. In contrast, however, writings from ancient Greece and Rome, such as the fables, have been deeply influential in depictions of donkeys as servile, stubborn and stupid. These opposing images are especially prevalent in Western culture because they derive from the deep contradictions within its very foundations: an underpinning Graeco-Roman philosophy and Jewish/Christian traditions. The literary and artistic texts used here exemplify these two opposing traditions and illustrate their influence on subsequent representations of donkeys down the centuries. Also evident are the emerging philosophies that recognize that animals be appreciated for what they are rather than for their utility to humans. How we perceive and represent animals will ultimately affect our treatment of them.

Although donkeys do not feature in high Greek literature (unlike horses), they are a common feature of popular proverbs and fables, serving in such moral tales as allegories of human weakness and folly. The animals are used as human exemplars; these writings are not concerned with the actual animal. Because they were a common feature of everyday life, donkeys' characteristics were readily recognizable for stories of moral instruction. Aesop (*c.* 620–564 BC) alone wrote around twenty fables in which donkeys are invariably depicted as foolish. In *The Ass in the Lion's Skin*, for example, a donkey is very proud of himself when he puts on a lion's skin that had been left out to dry by some hunters. All flee at his approach and he feels very

Sebastian Brant, 'The Fool with the Ass on his Back', from *Ship of Fools*, woodcut, 1494.

important. In his delight he lifts his head and brays; everyone then recognizes him and his owner gives him a sound beating. The moral is clear: not even a fine disguise will hide a fool. It is largely from these and similar fables that the donkey has its reputation for stupidity.

Probably the most influential story in determining attitudes to the donkey, however, was *The Golden Ass* (AD 160) by Apuleius.[2]

In Aesop's 'Fable of the Horse and the Ass', the lowly donkey finds that it is not always preferable to be a magnificent charger.

A mule by a water-mill, from a 13th-century bestiary.

It recounts the adventures of a young Greek called Lucius. His dabbling in magic resulted in him accidentally being transformed into a donkey, not an owl as he had wished. As a beast of burden, Lucius is condemned to serve humans, to live closely with them, to carry their burdens and to be abused by them. The donkey's commonly accepted traits such as stubbornness, foolishness, lustfulness and curiosity are used in the narrative to propel the story from episode to episode as Lucius is passed from one unpleasant character to another, who abuse and mistreat him in various unsavoury and often cruel ways. Lucius as the donkey is curious about human attributes, and he is grateful that his long ears enable him to eavesdrop from a distance. He is also pleased by the size of his phallus and there are many vulgar and sexual episodes.[3] It is the goddess Isis who restores Lucius to human form and gives him back his dignity because, as we saw in chapter Two, she hated donkeys due to their association with her enemy Seth-Typhon. Followers of Isis, the goddess of reason and logic, regarded the ass as impure and demonic, the cause of disorder and chaos. The donkey had come to represent all that was foolish, lustful and wicked. *The Golden Ass* was to have a significant impact on the way donkeys were perceived and depicted in literature over the ensuing centuries.

In the Middle Ages, bestiaries – compendia of animals describing their natural history together with a moral tale – divided animals into categories that represented the functions they fulfilled in human society; there was little evidence of interest in the reality of an animal's existence. The bestiary would describe an animal but then might use it as the basis for an allegorical tale. The animals were socially constructed as either 'good' or 'bad' and served to teach morality, often the embodiment of God's divine purpose. Asses were typically presented as melancholy, heavy, dull, clumsy and slow-witted. By the sixteenth century, Shakespeare in particular frequently encouraged the derogatory use of 'ass' by using it in his plays to describe stupid or clownish figures. This commonly used insult denoted an ignorant fellow, a perverse or conceited fool, and its use would immediately give rise to laughter.[4]

Most famously, perhaps, Shakespeare reworked a scene from *The Golden Ass* in *A Midsummer Night's Dream* (published in 1600), with Bottom's metamorphosis into an ass, except that Bottom is an unintelligent yokel, 'a rude mechanical', not an educated Greek gentleman. He is far from the sexual adventurer that Lucius was; he might be 'mortally gross' but he is not lascivious. As a human, Bottom is a likeable clown, 'a conceited dolt' who is over-enthusiastic, egotistical and often ridiculous; as an ass he has more down-to-earth, if uninspiring qualities. Here is the dull and clumsy ass of Aesop's fables and the bestiaries, lacking the sensitivities to appreciate the fantastical and magical world he has entered. As a stolid and unimaginative ass he is impervious to the advances of the beautiful Titania, Queen of the Fairies. The donkey is fit for everyday drudgery; he does not

Sancho Panza, and donkey, join Don Quixote in this commemorative statue in the Plaza de España, Madrid.

possess the inspired soul of the human that would enable him to enjoy the finer things in life.

Perhaps the most famous novel to feature a donkey as both animal companion and allegory for human nature is *Don Quixote* (1605), in which the horse and donkey are both metaphors for the men that ride them. Cervantes reveals a more positive attitude towards donkeys, perhaps because they had long been an important aspect of life in Spain. The Don's old horse Rosinante and Dapple, Sancho Panza's beloved donkey, are both true companions and mentors to their respective masters, as well as reflections of their characters.[5] As Antonio Vieira claims, the comparative psychology of the horse and donkey are faithfully portrayed in *Don Quixote*: 'the arrogance and caprice of the horse and the humble steadiness of the donkey'.[6] Sancho's loyalty to the Don is mirrored by his relationship with Dapple; they are good companions who enjoy nothing more than each other's friendship. Dapple brings stability and security to his life and Sancho is at his happiest and most peaceful when he is riding his donkey. When they are parted, Sancho's joy at their reunion is evident:

> Sancho ran immediately to his ass, and embraced him: 'How hast thou done?' cried he, 'since I saw thee, my darling and treasure, my dear Dapple, the delight of my eyes, and my dearest companion!' And then he stroked and slabbered him with kisses. The ass, for his part, was as silent as could be.

Both are simple and honest souls, to be admired for their loyalty and steadfastness. The artist Honoré Daumier was fascinated with the romantic tradition of Don Quixote, knight-errant vainly trying to live his dream of noble deeds, contrasted with the dumpy realist Sancho Panza. In his painting *Don Quixote and Sancho Panza* (*c.* 1860s) the mock-hero dashes off into the midday heat while Sancho sits, still and silent, on his donkey. There are obvious resonances of Christ on the donkey, as a symbol of meekness and humility. Dapple is not abused or denigrated but appreciated for his steadfast donkey qualities of loyalty and patience. Here it could be argued that something of the 'true' nature of donkeys is starting to be revealed, albeit within an allegorical context.

Honoré Daumier, *Don Quixote and Sancho Panza*, c. 1860s, oil on canvas.

Although an interest in animals and their welfare had been cause for debate amongst intellectuals in Europe for many centuries, the everyday plight of animals generally did not begin to improve until the latter part of the nineteenth century, at a time

R. H. Stoddard, a chained jenny and her foal illustrating Coleridge's poem 'To a Young Ass', engraving, *c.* 1880s.

when social change was an important aspect of the political agenda.[7] Romanticism, a European intellectual movement originating in the second half of the eighteenth century, influenced the ways in which humans viewed animals. Their identification moved from 'brutes' and 'beasts' to fellow creatures and, as they were to St Francis of Assisi in the thirteenth century, even companions and brothers.[8] In his poem 'To a Young Ass'

(1794), Samuel Taylor Coleridge highlights the contrasting attitudes of the educated elite about the ways in which donkeys were exploited and abused in Britain at the time, having been moved by the plight of a donkey foal he encountered on the grass at Jesus College, Cambridge, and feeling a compassion for him and a kinship with him. In the poem he addresses a foal with the words 'Poor little foal of an oppressed race!' and focuses on the misery of the creature and its chained and starving mother.[9]

Romanticism encompassed a compassionate, if idealized, view of nature, as well as a fascination with animals as both forces of nature and metaphors of human nature. In his narrative poem 'Peter Bell' (1819), William Wordsworth portrayed the donkey in a positive rather than ridiculous light. The poem is based on an incident from a newspaper, in which a donkey was found beside a canal, standing guard over the body of his dead master who had fallen in. Wordsworth threads this anecdote into his long tale, emphasizing the loyalty of the donkey and his determination to stay with his master, despite being beaten and abused by Peter Bell, the man who finds the lone donkey and decides to steal him. He is a lawless man, indifferent to the beauties of nature; but after he mounts the donkey and is taken by him to the drowned man's widow's cottage, he is reformed:

> He lifts his head and sees the Ass
> Yet standing in the clear moonshine;
> 'When shall I be as good as thou?
> Oh! Would, poor beast, that I had now
> A heart but half as good as thine!'

Clearly Peter Bell's journey on the donkey has Christian significance, as does the representation of the donkey as loyal and good;

but an appreciation of donkeys as actual animals rather than as representations of human nature is becoming more apparent.

Companionship and appreciation of her steadfast qualities are not in the forefront of Robert Louis Stevenson's mind as he describes the journey he made through the Cevennes in southern France with a little mouse-grey donkey called Modestine in his book *Travels with a Donkey* (1879).[10] He constantly loses patience with their slow progress and he hits her to try to make her go faster. He has already loaded her up with a very heavy bundle and, losing patience, he hits her twice across the face. Her response, however, makes him stop and think: 'It was pitiful to see her lift up her head with shut eyes, as if waiting for another blow. I came very near to crying.' He lightens her load – but continues to beat her and prick her flesh with a sharp goad. It is only when he has sold her and he is alone that he appreciates what she meant to him. 'I had lost Modestine. Up to that moment I had thought I hated her, but now she was gone, "And oh the difference to me!"' Stevenson's treatment of Modestine reflects the more usual and generally accepted attitude towards donkeys and, although he realizes what he has lost once she has gone, it is in regard to his own loss, rather than an appreciation of his donkey companion.

A totally different attitude is evident in Juan Ramón Jiménez's relationship with his donkey companion in *Platero and I* (1914).[11] On Jiménez's travels through Andalusia, Platero is treated as an equal and a friend. This is indeed a contemporary interpretation of human and animal interrelationships and it mirrors and reflects, albeit in a fictional and allegorical context, debates about the superiority of humans and the lowly inferiority of donkeys. Jiménez's donkey is his companion and confidant with whom he has long, philosophical conversations. This is the human and non-human animal bond at its most serene. Jiménez

Walter Crane, frontispiece to R. L. Stevenson's *Travels with a Donkey in the Cevennes* (1907).

grieves for all the donkeys that have been badly used and is distressed by the fate of all the neglected donkeys that they come across on their travels, such as the ancient donkey who has been abandoned at the rubbish tip where 'he will freeze to death in this high ravine, pierced by the north wind'. Jiménez does not see the difference in status between himself and Platero; when he takes his friend to visit a beautiful orchard and the guard at the entrance tells him that he may not enter with the donkey, Jiménez is surprised: 'What donkey?' When he realizes that the guard means his friend Platero they go on their way without visiting the orchard. It is an expression of human friendship bestowed on a donkey: after all, we cannot know if Platero feels as his master does about their relationship.

Perhaps the most famous of all literary donkeys in the West, especially among children, is Eeyore in A. A. Milne's *Winnie the Pooh* (1926).[12] As we have seen, donkeys as dependable companions form a popular thread in modern Western literature, although they are also invariably used as human stereotypes. In this book, the animals/toys are all representations of certain types of adults: Eeyore is the melancholy, grumbling old relative. In the bucolic and idealized setting of Hundred Acre Wood, Eeyore stands alone, quiet and introspective:

> The old grey donkey, Eeyore, stood by himself in a thistly corner of the Forest, his front feet well apart, his head on one side and thought about things. Sometimes he thought to himself sadly 'why' and sometimes he thought 'wherefore' . . . and sometimes he didn't quite know what he was thinking.

He can be a good friend but he is most often self-absorbed and glum. Eeyore echoes the gloomy, stolid and unimaginative asses of the bestiaries. He does not ask questions of his companions but tells them that everything 'is all the same to me'. As from Homer's *Iliad* through to Modestine, Eeyore exemplifies an innate understanding of donkeys' mundane lives, often of suffering, and the stoic resignation to accept their fate.

Benjamin, the old donkey in George Orwell's *Animal Farm* (1945), also demonstrates these qualities.[13] Once again, Benjamin is companion and loyal friend, here to Boxer, the old carthorse. He is a wise old donkey who shows little emotion. He is not taken in by the revolution but remains true to himself and to his old friend. Benjamin knows nothing will really change – revolutions will come and go but life will continue as before, harsh as it always was: 'things never had been, nor ever could be much better or much worse – hunger, hardship, and disappointment being, so he said, the unalterable law of life'. Benjamin is, of course, yet another human exemplar but his characteristics are to be admired rather than reviled; and to those who appreciate donkeys, his attributes of common sense, loyalty and wisdom, along with acceptance of his fate, certainly ring true. He can also be taciturn and stubborn but he has the wisdom to take a long-term, more mature vision than the other animals.

A more recent example of a 'companion donkey' in stories for children has quite a different character to those described so far, apart perhaps from the foolishness of the fables themselves. He is the ebullient Donkey in the film *Shrek* (2001).[14] Donkey, his very lack of name implying his relative unimportance, is both a clown and servant; homeless, he has to work hard as the servile travelling companion in order to maintain his lowly place with the other characters.[15] Donkey is depicted as less civilized than the other characters (even if they are green ogres): he feeds in the meadows and rolls in the dirt, and he is the only character to express sexual fantasies, as he discovers inter-species sex

The ebullient Donkey from Andrew Adamson's and Vicky Jenson's film *Shrek* (2001) may be full of fun and nonsense but he also knows the sadness of his kind.

when he seduces the female dragon. Donkey is full of cheek and nonsense while Benjamin and Eeyore are gloomy, thoughtful and quiet; but all learn how to survive in the environments in which they find themselves. They are still represented as having human characteristics but all display donkey 'behaviour' as well. They can be steadfast friends, once their confidence has been earned – and all know what it is to be ridiculed and to feel lonely and outcast.

The film *Au Hasard Balthazar* (1966) encapsulates the dichoomy between the actual and symbolic in the literary and visual representations of donkeys.[16] The director Robert Bresson has a profound understanding and appreciation of donkeys, of how they are represented and how that affects their treatment. In this moving and powerful black-and-white film, the protagonist, Balthazar, is presented as both a real and suffering animal and as a symbol of purity and humility. Bresson has said that he thinks the donkey 'the most important, the most sensitive, the most intelligent, the most thoughtful, the most suffering of animals'. Balthazar is used, abused and neglected by the inadequate, sometimes cruel, humans around him, yet remains steadfast and enduring, accepting of his fate and his punishments.

The film releases a profusion of symbolism associated with donkeys, especially from Christian imagery. It connects the corporeal with the spiritual in the innocent, Christ-like figure of the abused Balthazar. This is especially powerful in the closing moments of the film as the battered old donkey lies down to die, bleeding from gunshot wounds – a lonely death in an Alpine meadow encircled by a flock of wandering sheep. We are left in little doubt as to the religious symbolism of the role of Balthazar, since he is baptized in the opening scenes, is crowned with a wreath of wild flowers, carries human burdens and is bound, harnessed and whipped by them. However, while the biblical associations and allusions in the film are many, Balthazar is also depicted as a real donkey that experiences loss, pain and death. The final shot of his dead body reinforces this harsh reality.

As a narrative, *Au Hasard Balthazar* borrows from a number of literary traditions. It is inspired in parts by Apuleius' *The Golden Ass* and by Fyodor Dostoevsky's *The Idiot* (1868–9). The structure of Balthazar's story is similar to that of *The Golden Ass* since he is passed from person to person, receiving abuse

In Robert Bresson's film *Au Hazard Balthazar* (1966) the abandoned Balthazar, bearing the burdens of his mistreatment by humans, lies down to die in a meadow surrounded by sheep.

in return for his loyal service. However, Balthazar himself is similar to Dostoevsky's Prince Myshkin, the Christ-like protagonist of *The Idiot*. Pure and good, Balthazar suffers and endures, a silent witness to human folly and cruelty. He meekly accepts his fate because he must; he has no control over what happens to him. Bresson has said that he is simply a donkey who

> walks or waits, regarding everything with the clarity of a donkey who knows it is a beast of burden, and that its life consists of either bearing or not bearing, of feeling pain or not feeling pain, or even feeling pleasure. All of these things are equally beyond its control.[17]

Rather than the animal being used to represent human failings, it is the humans in this tale who embody all the vices. Bresson says the film is about humanity's anxieties and desires when faced with an animal who 'is completely humble, completely holy'. Bresson has taken the slurs heaped on the donkey in the *The Golden Ass* and has turned the tables on human notions of their own superiority. It is the people in this story who are driven by pride, greed and lust; the donkey is innocent and incorrupt.

In *The Wisdom of Donkeys* (2008), Andy Merrifield contemplates 'real' donkeys, along with representations of them, as he travels through the French countryside with a donkey companion.[18] This contemporary text illustrates the changing philosophical attitude towards animals in that the author specifically sets out to understand and appreciate donkeys in their own right. Merrifield is an adventurer and philosopher who, unlike Stevenson, learns the art of donkey walking as he and Gribouille, a gentle chocolate-brown donkey, amble through the picturesque Auvergne region. The friend from whom he borrows Gribouille explains: 'donkeys make you patient, calm, tranquil . . . and they have a lot of tenderness'. As Merrifield meanders at Gribouille's pace, he learns such patience and he reflects on the meaning of life and on donkeys past and present. He talks to his donkey companion as Jiménez talked to Platero and calls him 'the great philosopher' and 'a profound presence'. After spending hours observing Gribouille, Merrifield is impressed by his sensitivity and intelligence. He says he will not forget the 'innocent gaze' of a donkey's eyes, 'the gravest and most reasonable eyes the world has seen'.

> Behind that calm patience, beyond the tranquillity, here is an animal who has been through rough times, who has suffered and knows about tragedy and pain . . . Donkeys know about moments of joy and happiness, about finding the perfect dandelion and a quiet meadow in the sun . . . but they know, too, how cudgels and goads abound everywhere . . . Always.[19]

Our changing attitudes towards donkeys are also represented in the visual arts. Although images may differ they all represent the close connection between humans and donkeys over time and place. The first narratives about our lives together, before the written word, were cave paintings: human attempts not only to record their world but also to make sense of it and their place within it. In visual representations over the centuries, donkeys have been portrayed being hunted for food, as beasts of burden, as companions, even as gods to be worshipped. They have been a constant source of inspiration and myth making for storytellers and artists of different cultures.[20]

It is within the context of their interaction with humans that animals have largely been depicted: scenes within which only animals appear in a naturalistic landscape were extremely rare.

For instance, in hunting scenes in which wild asses appear they are man's victims; in domestic scenes they are depicted as his helpers. The palace walls of ancient civilizations of, for example, Babylon and Assyria were covered in animal forms in these wild and domestic roles. Roman and early medieval mosaics also show detailed observation and a close interest in animal life.[21] A fifth-century Byzantine mosaic in Istanbul depicts a boy feeding his donkey after a day's work. However, although carefully drawn, the donkey has a human expression: looking outwards, turning his head away from the proffered food, the donkey looks sad and tired.[22] Such drawings often revealed not only close observation of the animal but also its significance to humans. Donkeys appear often in Western medieval art not only because of their importance in the day-to-day lives of the people, but also because of their significant religious symbolism.

One of the earliest sculptures of Christ on a donkey appears on a Roman sarcophagus from the fourth century AD, in the Museo Nazionale delle Terme in Rome. The humble donkey is shown as small and downtrodden as she plods along, head down, big ears laid back, bowed under the weight, with her foal under her belly. Jesus' entry into Jerusalem on the back of a donkey is prominent in early Christian art and the event is depicted on numerous Roman sarcophagi. These images are clearly intended to contrast with those scenes depicting conquering Roman emperors entering a city in ostentatious displays, mounted on a horse or riding in a horse-drawn chariot. The donkey was an indication both of his status and purpose, highlighting the humility and meekness of Jesus. Art historian Thomas Mathews proposes:

> The importance of the ass in Early Christian art signals a new attitude towards the whole animal kingdom. While the Classical world sometimes drew moral lessons from

animal behaviour and made them act out human dramas in Aesop's fables, the Christian mind saw them as somehow collaborators in human endeavours . . .[23]

However, in early Christian art, the donkey represents both the differentiation of the newly developing religion from official Roman religion and, simultaneously, the borrowing of images from classical traditions by some of the first Christian artists. The donkey was a common feature in art depicting Dionysus and related priapic gods and followers.[24] The processional scene of Jesus entering Jerusalem on the back of a donkey already carried a complicated melding of traditions.

Some of the most famous of these images in Gothic art are to be found in the numerous depictions of the *Nativity* and the *Adoration of the Magi*, virtually all including the donkey as an important symbolic element.[25] In Gentile da Fabriano's predella panel of the *Nativity* (1423) the donkey is in the centre of the scene, kneeling down on the crib so that he can better see the baby Jesus lying on the ground. In that by Domenico Ghirlandaio

Dionysus, the Greek god of wine and fertility, was often depicted riding on a donkey, also a symbol of fertility and lust; here a marble statue from the 1st century BC.

(1492) there is a beautifully observed, grey donkey with its ears pricked forward, giving his full attention to the baby. Gentile da Fabriano's beautiful and elaborate *Adoration of the Magi* (1423) is crowded with incident, with humans and animals, yet the focus is on the central group, which includes a sympathetic depiction of the ox and ass. The triptych altarpiece of the *Adoration of the Magi* (*c.* 1495) by Hieronymus Bosch, also known as the *The Epiphany* and now in the Museo del Prado in Madrid, depicts the traditional scene in an untraditional manner. Mary sits with the baby on her lap outside a decaying old stable in Bethlehem. In the centre of the picture, a donkey's head is framed in the darkened doorway: there is no sign of the ox, which is very unusual. The donkey's head could be a reference to the calumny of ass worship laid on the Jews: the donkey, although appearing perfectly benign, could well have darker implications in this painting (as described in chapter Two).

As well as being depicted for symbolic purposes, donkeys appear as an important element of the narrative of a scene, invariably as a means of transportation, especially in those numerous works of art depicting the flight of the Holy Family to safety in Egypt, or those of Jesus entering Jerusalem. Virtually all of these include a donkey, usually in a position signifying its importance. The *Flight into Egypt* by Giotto di Bondone (1304–6) in the Scrovegni Chapel in Padua, for example, has the donkey centre stage and realistically portrayed. The shape of the head, the droop of the neck and the gentle eye reveal a careful study of and suggest a fondness for donkeys. Two of the most significant depictions of Christ entering Jerusalem are a panel of Duccio's altarpiece (1308–11) for Siena Cathedral and that by Giotto in the Scrovegni Chapel. In both, the donkey is central to the event and is depicted carefully. In the former, a foal accompanies its mother; in the latter, the donkey is drawn simply, powerfully and with empathy.

Giotto's fresco of the *Nativity* in the Scrovegni Chapel, Padua, 1305.

Donkeys continued to play their role in religious art in the Renaissance, sometimes as beasts of burden, at others in a symbolic role. Donatello's fifteenth-century relief on the high altar of Sant'Antonio in Padua depicts a donkey kneeling before the Blessed Sacrament, perhaps leading the way to the veneration of Christ but certainly a central element in the ritual. The painting of the *Ecstasy of St Francis* by Giovanni Bellini (*c*. 1480), now in the Frick Collection, New York, is interesting as there is a little donkey in the background of this grand scene of St Francis in ecstasy outside the opening of a large cave. Perhaps the donkey symbolizes the great humility of St Francis's inner self, while also representing the close connection he had with his actual donkey.

Giotto's fresco of the *Flight into Egypt* in the Scrovegni Chapel, Padua, 1304–6.

It has been said that on his deathbed St Francis thanked his donkey for carrying him, while the donkey wept. Another popular topic for religious paintings was the story of the Flood and Noah's ark. In several paintings depicting this event, the donkey features as the domestic beast of burden rather than as one of a pair entering the ark. In Castiglione's *In Front of Noah's Ark* (*c.* 1650), the donkey, bearing Noah's belongings, is prominently displayed as an important element in the centre of the scene. In Jan van Kessel's *Boarding the Ark* (1725), however, there is a magnificent horse dominating the foreground of the picture, head held high, nostrils flaring and mane flowing. In the background of the painting, a lowly donkey is weighed down by Noah's possessions.

Albrecht Dürer, *The Flight into Egypt*, oil on panel, 1496.

In the eighteenth and nineteenth centuries changing attitudes are especially apparent in Western visual art as people moved from agricultural to industrial societies. The set-piece religious scenes of the past were replaced by a wide variety of themes and styles and, in many, animals moved from supporting roles to centre stage. As we have seen, growing awareness of and concern for the welfare of animals was becoming a major social issue. Political and social reform resulted in overworked, abused and half-starved animals being depicted, the pictures capturing the suffering of animal industrial labourers. There are engravings and paintings featuring neglected and starving donkeys, reminiscent of Coleridge's poem 'To a Young Ass'.[26] This was also a time of religious upheaval when the concept of evolution challenged old beliefs, and attitudes towards animals were multiple and

In this Giotto fresco in the Scrovegni Chapel, the realistically drawn donkey takes centre stage as he bears Christ towards the Golden Gates of Jerusalem.

In this 1480s oil by Giovanni Bellini, St Francis prays while his donkey looks on in the background.

complex. Another influence was Darwin's theory of evolution, which encouraged a more scientific approach to animal representation, leading to carefully observed and minutely detailed pictures of animals. However, as attitudes changed towards animals, especially amongst the elite, so they came to be depicted differently: there was a greater recognition that they needed to be treated and represented with dignity.

The emotional expressions of animals became a significant subject for nineteenth-century Romantic artists, who would begin to break down the traditional distinctions between history painting and animal art. Many paintings depicted animals in noble settings, as being equal to the human subject, an obvious example being majestic horses bearing conquering heroes in times of war. Paul Delaroche's *Napoleon Crossing the Alps* (1848–50) is unusual, although realistic, in that Napoleon is riding a

Gustave Courbet depicted the harshness of rural life, including this donkey painted in 1863.

mule as he leads his troops across the treacherous mountain pass. The mule is obviously undernourished and weary from its struggle over the Alps in the icy winds and harsh conditions, and the painting captures the difficulties for both human and non-human animals. Delaroche painted this picture as a contrast to the famous series of paintings of Napoleon's journey by Jacques-Louis David (1801–5). These are idealized and romanticized, and Napoleon is depicted, quite incorrectly, riding on a magnificent charger.

Edwin Landseer (1802–1873) probably best typifies the Romantic representation of animals.[27] His paintings, in which carefully observed animals display emotions, became very popular. His love for animals, especially domestic companions, touched people's hearts and brought art to those who otherwise knew little of it – at a time when they had lost contact with working animals and so were developing closer relationships with their pets. Landseer's fondness for donkeys is evident in the many sketches, etchings and paintings he made of them.

Paul Delaroche, *Napoleon Crossing the Alps*, oil on canvas, 1850.

In *Shoeing* (1844) an attractive bay mare is being shod while her faithful companions, a donkey and a bloodhound, look on. The mare is the obvious centre of attention and her glossy coat shines out of the picture, in contrast to that of the shaggy donkey. The aristocratic mare turns her powerful neck towards the humble donkey; she permits the relationship with grace while the donkey bends his head in a deprecating manner. However, the donkey is endearing as he waits patiently, meekly bearing his saddle yet with the flamboyant flourish of a red poppy in his bridle. He is not a downtrodden or abused beast of burden; he is a contented donkey, an appreciated member of the group.

Although often labelled sentimental, many of Landseer's paintings also typically made a social comment. In *Boy, Donkey and Foal – Mischief in Full Play* (1822), for example, he makes a strong statement about the general perception and treatment of donkeys. The picture could be illustrating Bartholomew Anglicus's belief that although donkeys are born fair and comely, as they grow older they 'ever waxeth fouler', as the appealing beauty of the foal in the scene is contrasted with the miserable appearance of the adult donkey. However, we are reminded of the probable reasons for this: unsuitable climatic and environmental conditions and abusive treatment. Despite being beaten by the boy on her back, this unhappy donkey cannot move forward because she is hobbled with a manacle and chain.

Thomas Sidney Cooper (1803–1902) had a similar passion for British domestic animals and produced many romantic, bucolic scenes of cows, sheep, horses and donkeys in the English countryside. Some of those featuring donkeys include *A Donkey on a Beach* (1879) and *A Donkey and Sheep in a Meadow* (1880): all are detailed depictions capturing the gentle, patient, if sometimes downtrodden, nature of these animals. Donkeys continue

Constant Troyon (1810–1856), *The Shepherd's Rest*, oil on panel.

Henri de Toulouse-Lautrec, *Raoul Tapié of Celeyran on a Donkey*, 1881, oil on board.

to appear as appealing domestic animals in farmyard scenes in a variety of styles from the late nineteenth century into the twentieth, such as *Farmyard with Donkeys and Roosters* (*c.* 1880) by Adolphe-Joseph-Thomas Monticelli and *Farmyard Friends* (1921) by Edgar Hunt. *The Vegetable Garden with Donkey* (1918) by Joan Miró is a detailed drawing from rural life in Motroig, which vividly encapsulates those long summer days of childhood. The brilliant, golden colours reveal the magic of childhood memories and the longing for the simplicity of rural life, including the humble family donkey, the only figure in this idealized landscape.

Animals continued to grow in popularity as subjects for the visual arts in the twentieth century, in scientific, aesthetic and symbolic terms. Even though an appreciation of the beauty of the animal in its own right and our growing understanding of the true natures of animals make the symbolic approach less relevant, it remains evident in many paintings. The *Blue Donkey* (1925) by Marc Chagall is a blend of fantasy, nostalgia, folklore and religious imagery. A passion for life and colour are more important than realistic depictions in this deceptively simplistic painting. Yet this whimsical bright blue donkey, with religious connotations, reminds of happy childhood memories and a concomitant fondness for donkeys. The close connection between children and donkeys makes an appealing subject for many artists' work, most famously perhaps Pablo Picasso's painting of his son on his pet donkey: *Paolo on a Donkey* (1923).

Marc's frieze is inspired by the ancient Egyptian wall frieze of donkeys (*c*. 2700–2600 BC) in the Rijksmuseum, Leiden, 'thus bringing together one of the oldest depictions of donkeys as beasts of burden and a modern interpretation'.[28] Marc's main interest was in the dynamic connection between animals, humans and the natural world they shared, which he explored through radical compositions and bold experiments with colour. In attempting to convey the spirit and inner being of his subjects, Marc often depicted the animals in non-naturalistic colours. He associated blue with masculinity and considered it to be the most deeply spiritual of the colours he used. His powerful donkeys are proud and free, content in their herd where they belong: they are not downtrodden, individual beasts of burden. It is this kind of attempt to understand an animal in its own right that will lead to a greater appreciation of

Inspired by ancient Egyptian wall friezes of working donkeys, Franz Marc's *Donkey Frieze*, 1911, is filled with a herd of proud and free blue donkeys.

what it means to be that animal and to a more compassionate outlook. In the words of Mahmoud Darwish:

> The best spectator on the world stage is the ass
> A peaceful, wise animal that feigns stupidity
> But he's patient, smarter than we are
> In the cool, calm way that he watches
> The making and progress of history
> Armies march past him and flags change
> Along with the birds painted on them
> While he watches unchanged.[29]

Timeline of the Donkey

8000 BC

Different species of wild ass ranged across areas of North Africa and Asia. They were hunted for food and for sport.

4500 BC

The earliest long-distance merchant trading routes are formed using donkeys

4000 BC

The North African wild ass *Equus africanus* is domesticated. They are kept in large herds for their milk and meat.

4000 BC

Sumerians are using donkeys to pull carts

3000 BC

Ancient Egyptians are using donkeys as their only beast of burden

***c.* 300 BC**

Donkeys introduced into China

***c.* 200 BC**

The Greeks bring their donkeys with their vines to their colonies all along the coast of the Mediterranean

***c.* 50 BC**

Cleopatra bathes in asses' milk

***c.* AD 30**

Jesus rides a donkey into Jerusalem on Palm Sunday

AD 50

Donkeys and mules travel to Britain with the Roman invasion

1784

Spanish donkey jacks are imported into USA by George Washington to breed mules

1793

The first few donkeys arrive in Australia

1822

A donkey appears in court in London as evidence in the first conviction for cruelty to animals (Martin's Act)

1915

Simpson and the donkey carry wounded soldiers at Gallipoli in the First World War

2800 BC	2600–2400 BC	1500 BC	1000 BC	AD 100
Sumerians breed the first hybrids, onager and donkey crosses	Donkeys pull Sumerian war chariots	Donkeys accompany the Israelites on their journey into the desert	Donkey a common means of transport throughout Egypt and Asia	Mule breeding a huge industry in ancient Rome – mules and donkeys used as transport in the expansion of the Roman Empire

AD 160	1200	1423	1495	1500
Apuleius writes *The Golden Ass*	St Francis has a special bond with his donkey	*The Nativity* is painted by Gentile da Fabriano	The first donkeys arrive in America with Columbus	Mules used in Europe by high-ranking ecclesiastics such as Cardinal Wolsey

1929	1940s	1969	1971	2005	2009
Miniature donkeys imported into the USA by Robert Green	Donkeys culled in the USA, Australia and South Africa	The Donkey Sanctuary established in the UK	Wild Horses and Burros Act passed in USA	Donkeys working on Blackpool beach in the UK to have lunch breaks	Thousands of feral donkeys slaughtered in Australia

References

INTRODUCTION

1 P. A. Vieira, 'Our Brother the Donkey', in *Kinship with Animals*, ed. K. Solisti and M. Tobias (San Francisco, CA, 2006), p. 134.
2 A. Beja-Pereiraet et al., 'African Origins of the Domestic Donkey', *Science*, CCCIV (2004), p. 1781.
3 V. Morrell, 'Cruelest Place on Earth: Africa's Danakil Desert', *National Geographic* (October 2005), pp. 45–52.
4 D. Fielding and R. Pearson, eds, *Donkeys, Mules and Horses in Tropical Agriculture Development* (Edinburgh, 1991).
5 J. P. Mahaffy, 'On the Introduction of the Ass as a Beast of Burden into Ireland', *Proceedings of the Royal Irish Academy*, XVIII (1917), pp. 530–38.
6 In fact, 'ass' (from the Latin *asinus*) is totally unrelated to arse (from Old English), spelt 'ears' but pronounced more like eh-arss. The two became confused due to the vowel shift from 'ah' to short 'a' (as in 'dance'), plus the salacious thrill of the 'naughty word'.
7 F. Zeuner, *A History of Domesticated Animals* (London, 1963), p. 378.
8 Comte de Buffon, *Histoire naturelle, générale et particulière* (Paris, 1749–88).
9 K. Marshall and Z. Ali, 'Gender Issues in Donkey Use in Rural Ethiopia', in *Donkeys, People and Development: A Resource Book of the Animal Traction Network for Eastern and Southern Africa (ATNESA)*, ed. D. Fielding and P. Starkey (Wageningen, Netherlands, 2004), pp. 64–8.
10 M. J. George, 'Riding the Donkey Backwards: Men as

Unacceptable Victims of Marital Violence', *Journal of Men's Studies*, III/2 (November 1994), pp. 137–59.

11 Buffon, *Histoire naturelle*.

12 A fine 'mane of hair' was admired in both women and horses at this time, and still is. It was noted in several Greek texts that a proud mare had her mane shaved off before she would allow a donkey jack to mount her in order to breed a mule.

13 J. Shelton, 'Valued and Reviled: Contrasting Images of Donkeys in Ancient Greek and Roman Authors', conference paper, at *The Role of the Donkey and Mule in the Culture of the Mediterranean'* (Hydra, 2007).

14 R. Borwick, *The Book of the Donkey* (London, 1981).

15 Quoted in W. Long, *Asses vs Jackasses* (Portland, OR, 1969), p. 6.

1 *EQUUS ASINAS*: ORIGINS, DOMESTICATION, BREEDS AND CHARACTERISTICS

1 Robert Green, responsible for importing miniature donkeys into the United States in 1929, is reported to have made this claim.

2 S. W. Baker, *Exploration of the Nile Tributaries of Abyssinia* (Hartford, CT, 1868).

3 P. Moehlman, F. Kebede and H. Yohannes, 'The African Wild Ass (*Equus africanus*): Conservation Status in the Horn of Africa', *Applied Animal Behaviour Science*, LX/2–3 (1998), pp. 115–24.

4 Kimura, B., et al., 'Ancient DNA from Nubian and Somali Wild Ass provides insight into Donkey Ancestry and Domestication', *Proceedings of the Royal Society* (28 July 2010).

5 M. Yousef, 'Physiological Responses of the Donkey to Heat Stress', in *Donkeys, Mules and Horses in Tropical Agriculture: Proceedings of colloquium held 3–6 September 1990, Edinburgh, Scotland*, ed. D. Fielding and R. Pearson (Edinburgh, 1991), p. 96.

6 A. Beja-Pereira et al., 'African Origins of the Domestic Donkey', *Science*, CCCIV (2004), p. 1781.

7 J. Clutton-Brock, ed., *The Walking Larder: Patterns of Domestication, Pastoralism and Predation* (London, 1989).

8 For a full discussion of the discovery and dating of ancient equid bones, see R. Meadow and H.-P. Uerpmann, eds, *Equids in the Ancient World* (Wiesbaden, 1986), vol. I; F. Marshall, 'Ethnoarchaeological Perspectives on the Domestication of the Donkey: Bone Commons Forum' (2006): http://www.alexandri-aarchive.org/icaz/icazForum/ (accessed 1 January 2008).

9 S. Rossel et al., 'Domestication of the Donkey: Timing, Processes and Indicators', *PNAS: Proceedings of the National Academy of Sciences of the USA*, CV/10 (2008), pp. 3715–20.

10 Domestic donkeys weigh between 80–480 kg and their height at the withers ranges from 80–160 cm. Their colours range from black to white but donkeys are predominantly varying shades of grey or brown. They can also be roan, piebald, skewbald or broken-coloured, that is, a combination of brown and white or black and white markings. More rarely they can be pure white or chestnut.

11 'The Donkey's Advocate' (2008): http://www.travelintelligence.com/travel-writing/1001331/Europe/Spain/Andalucia-C (accessed 26 February 2008).

12 Hellenic Communication Service, 'Greece's Donkeys Moving into History at a Rapid Pace' (2007): http://www.helleniccomserve.com/greecedonkeys.html (accessed 12 February 2009).

13 I. L. Mason, *A World Dictionary of Livestock Breeds, Types and Varieties*, 3rd edn (Wallingford, Oxon, 1988).

14 W. Kugler, H-P. Grunenfelder and E. Broxham, *Donkey Breeds in Europe: Inventory, Description, Need for Action, Conservation* (Basel, 2008)

15 E. Svendsen, *The Professional Handbook of the Donkey* (London, 1986).

16 According to Pliny the Elder in the first century AD (*Natural History*, Book 8, 46), each male wild ass is the lord of his own herd of females. Because he is jealous of rivals, he watches his females and castrates with a bite any male foals that are born. To prevent this, the females try to give birth in secret. The wild ass indulges in a great deal of sexual activity.

17 http://sciencestage.com/resources/rrpedia/African_Wild_Ass#Behavior.

18 However, in the domestic state, both male and female donkeys bray – and for a variety of reasons.

19 See, for instance, C. Hadley, 'Curbing Wily Coyote: Using Donkeys to Control Coyotes', *Sports Illustrated*, LXXV/2 (1991), pp. 5–9. B. Tapscott, 'Guidelines for Using Donkeys as Guard Animals with Sheep' (1997): http://www.omafra.gov.on.ca/english/livestock/sheep/facts/donkey2.htm (accessed 25 October 2004). D. Jenkins, 'Guard Animals for Livestock Protection: Existing and Potential Use in Australia', Vertebrate Pest Research Unit, Orange Agricultural Institute (Orange, 2003).

20 http://www.elisabethsvendsentrust.org.uk/view/help.

21 A. S. Aluja and F. Lopez, 'Donkeys in Mexico', in *Donkeys, Mules and Horses in Tropical Agricultural Development*, ed. D. Fielding and R. A. Pearson (Edinburgh, 1991), pp. 1–7. Donkeys have a life expectancy of 40 years; but the average life of the working donkey in most developing countries is rarely over twelve years and is as low as eight in many cases. A well-cared-for donkey living under appropriate conditions can live to 45 or 50 years of age. Although donkeys do not suffer from sickness in the way that horses do, contemporary studies have revealed that the greatest cause of illness and death in donkeys is parasitism.

22 See, for example, Shahabat Khan, 'Donkey Management and Utilisation in Peshawar, Pakistan', in *Donkeys, People and Development: A Resource Book of the Animal Traction Network for Eastern and Southern Africa (ATNESA)*, ed. D. Fielding and P. Starkey (Wageningen, Netherlands, 2004), pp. 236–7; and 'Brooke Hospital for Animals: Home Page' (n.d.): http://www.thebrooke.org/ (accessed 19 February 2006).

23 'Asinus: The Virtues of Donkey Milk' (n.d.) http://www.asinus.fr/histoire/info.html: (accessed 1 January 2007).

24 'Donkey Milk Aids Ill Woman', *Dungog Chronicle* (27 September 2006), p. 9.

25 'Asinus: Asinerie de feillet' (2007): http://www.asinus.fr/savon/info.html: (accessed 14 October 2007).

26 W. Thackeray, 'Respecting asses', *Cornhill Magazine* (January–June 1864), pp. 71–2, (p. 71).

27 A. Agangaet et al., 'Carcass Analysis and Meat Composition of the Donkey', *Pakistan Journal of Nutrition*, II/3 (2003), pp. 138–47.

28 Donkey trade deal 'may be worth $20m': http://www.abc.net.au/news/stories/2009/06/22/2605319.htm. Peter Morley, 'Aussie Donkeys to Boost Chinese Women's Libidos', *Courier Mail* (22 June 2009): http://www.news.com.au/couriermail/story/0,23739,25667159-3102,00.html (accessed 22 June 2009).

29 F. Zeuner, *A History of Domesticated Animals* (London, 1963).

30 J. Clutton-Brock, *Horse Power: A History of the Horse and the Donkey in Human Societies* (Cambridge, MA, 1992).

31 There have been reported cases of hinnies producing offspring, but these are rare and the validity of some reports has been questioned.

32 Old Red, however, a famous American mule, won many races against thoroughbreds, much to their owners' annoyance. See J. Baker, 'Old Red – Kentucky's Wonder Mule', *The Brayer*, II (1999), p. 18. A popular anecdote concerning the speed and endurance of a mule involves a race between General Custer on his horse and Bill Cody on his mule. The faster horse was exhausted by the distance over harsh terrain and collapsed and died, leaving the mule to win the race. In Australia, a similar challenge is said to have taken place. In this race, the horse actually won but was found dead the following morning, while the mule had recovered completely and was fresh for the next day's ride.

2 DONKEYS IN HUMAN HISTORY, MYTHOLOGY AND RELIGION

1 L. Hobgood-Oster, *Holy Dogs and Asses: Animals in the Christian Tradition* (Urbana and Chicago, IL, 2008), pp. 24–5.

2 A. Dent, *Donkey: The Story of the Ass from East to West* (London, 1972), p. 26.

3 A. Beja-Pereira et al., 'African Origins of the Domestic Donkey', *Science*, CCCIV (2004), p. 1781.

4 M. Griffith, 'Horse and Donkey Work: Equids and the Ancient Greek imagination', *Classical Philology*, CI/3 (2006), pp. 185–246.

5 H. Ritvo, *The Animal Estate: The English and Other Creatures in the Victorian Age* (Cambridge, MA, 1987).

6 R. Borwick, *The Book of the Donkey* (London, 1981), p. 20.

7 H. Mayhew, *London Labour and the London Poor* (London, 1851).

8 The Trial of 'Bill Burns', under the Martin's Act (1822).

9 BBC News, 'Donkeys "must have lunch breaks"' (2005): http://news.bbc.co.uk/1/hi/england/lancashire/4533391.stm (accessed 21 April 2007).

10 J. Serpell and E. Paul, 'Introduction', in *Animals and Human Society*, ed. A. Manning and J. Serpell (London and New York, 1994), p. xi.

11 Griffith, 'Horse and Donkey Work', p. 229.

12 S. Rossel et al., 'Domestication of the Donkey: Timing, Processes and Indicators', *PNAS: Proceedings of the National Academy of Sciences of the USA*, CV/10 (2008), pp. 3715–20.

13 F. Marshall, 'How Wild Asses Became Donkeys to the Pharaohs 2008', *New Scientist* (22 March 2008): http://www.newscientist.com/channel/life/mg19726474.700-how-wild-asses-became-donkeys-of-the-pharaohs.html.

14 Dent, *Donkey*, p. 39.

15 W. M. Thackeray, 'Respecting Asses', *Cornhill Magazine* (January–June 1864), pp. 71–2.

16 R. Bulliet, *Hunters, Herders and Hamburgers: The Past and Future of Human-Animal Relationships* (New York, 2005), p. 159.

17 Ibid., p. 151.

18 Robert Young, *Analytical Concordance to the Bible* (Grand Rapids, MI, 1982).

19 Khalid Sindawi, 'The Donkey of the Prophet in Shi'ite Tradition', *Al-Masaq*, XVIII/1 (2006), pp. 87–98.

20 Phosphor Mallam, *The Donkey Book* (London, 1935).

21 Herodotus, *Herodotus: The Histories*, trans. R. Wakefield (Oxford,

1998). For a full account of the donkey's bray, see F. Brookshier, *The Burro* (Norman, OK, 1974), chapter 2.

22 T. Lessing, 'Asses and Masses', *Daily Living Age* (November 1931), pp. 260–62.

3 DONKEYS AND MULES COLONIZE THE AMERICAS, AUSTRALIA AND SOUTH AFRICA

1 C. Lummis, 'Brother Burro', *Land of Sunshine*, IV/3 (1896), pp. 107–8.
2 See, for example, C. Lummis, *Flowers of Our Lost Romance* (New York, 1929).
3 F. Brookshier, *The Burro* (Norman, OK, 1974).
4 Ibid.
5 Frank Waters, *Midas of the Rockies* (Denver, CO, 1949).
6 Lummis, 'Brother Burro', pp. 107–8.
7 Cited in W. Long, *Asses vs Jackasses* (Portland, OR, 1969), p. 26. This account is unusual in that the prospector obviously found donkeys to be hardier than mules.
8 For example, M. Stamm, *The Mule Alternative* (Battle Mountain, NV, 1992).
9 Long, *Asses vs Jackasses*, p. 77.
10 Department of Agriculture, *History of the Jack Stock and Mules in Missouri* (Jefferson, NJ, 1924).
11 See, for example, S. Olsen, ed., *Horses Through Time* (Boulder, CO, 1996).
12 Ibid.
13 Ibid.
14 Long, *Asses vs Jackasses*, p. 81.
15 J. Kramer, *Animal Heroes: Military Mascots and Pets* (London, 1982).
16 Long, *Asses vs Jackasses*, p. 79.
17 A. Vernon, *The History and Romance of the Horse* (New York, 1946).
18 Ibid., p. 19.
19 J. Bough, 'Value to Vermin: The Donkey in Australia', PhD thesis, Newcastle University (Newcastle, 2008).

20 M. J. Kennedy, *Hauling the Loads: A History of Australia's Working Horses and Bullocks* (Melbourne, 1992), p. 26.
21 K. Burbidge, *The Donkey Drivers* (Adelaide, 1986), p. 4.
22 *A History of the Beltana Pastoral Company Ltd* (Adelaide, 1965), p. 56.
23 H. M. Barker, *Camels and the Outback*, new edn (Perth, 1995), p. 71.
24 *Pastoral Review* (16 May 1916).
25 *Statistical Register*, 1906–41, annual (Adelaide, 1920).
26 *Pastoral Review* (16 March 1931).
27 G. Wellard, 'Bushlore', *Donkey Digest* (September 1997), pp. 5–7.
28 'Donkey Trails in the Kimberley', *Walkabout* (1949), p. 42.
29 Ibid., p. 42.
30 Olsen, *Horses Through Time.*
31 Ibid., p. 31.
32 Society for Animal Protective Legislation, 'Bill Reintroduced to Protect Wild Horses from Slaughter' (2007): http://www.saplonline.org/w_horses_intro110.htm (accessed 21 April 2007).
33 A. Walker, *Australian Donkeys* (Victoria, 1973), p. 24.
34 Department of the Environment and Heritage, *Feral Horse (Equus Caballus) and Feral Donkey (Equus asinus)* (Canberra, 2004).
35 J. Vandenbeld, *Nature of Australia: A Portrait of the Island Continent* (Crows Nest, 1988).
36 Ibid.
37 'Kimberley Collars Judas Donkeys' (1999), http://savanna.ntu.edu.au/publications/savanna_links9/judas_donkeys.html (accessed 25 March 2005).
38 M. Everett, *Summary of Donkey Control in the West Kimberley: Pastoral Memo – Northern Pastoral Region* (Derby, 2007).
39 T. Flannery, *The Future Eaters: An Ecological History of the Australian Lands and People* (Melbourne, 1994), p. 369.
40 Nancy Jacobs, 'The Great Bophuthatswana Donkey Massacre: Discourse on the Ass and Politics of Class and Grass', *American Historical Review*, CIX (2001), pp. 485–507.
41 P. Starkey, 'The Donkey in South Africa: Myths and

Misconceptions', in *Animal traction in South Africa: Empowering rural communities*, ed. P. Starkey (Gauteng, 1995), pp. 139–51.

42 P. Jones, 'Response to Demand: Meeting Farmers' Need for Donkeys in South Africa', in *Donkeys, People and Development: A Resource Book of the Animal Traction Network for Eastern and Southern Africa (ATNESA)*, ed. D. Fielding and P. Starkey (Wageningen, Netherlands, 2004), pp. 198–202.

43 N. Jacobs, *Environment, Power and Injustice* (Cambridge, 2003), p. 216.

44 Ibid., p. 174.

45 Later donkeys were even blamed for causing traffic accidents; they failed to get out of the way of cars.

46 Jacobs, *Environment, Power and Injustice*, p. 201.

47 Ibid., p. 214. In Australia, when they were no longer useful to Europeans, donkeys were associated with Aborigines living on the missions, where they were ridden by children and used for collecting firewood by adults. Now that they are slaughtered as vermin, many Aborigines do not recognize this European status and see the donkey as 'belonging' to the land because they have lived there for so long. They also remember the biblical associations of the donkey from the stories they were taught on the missions and do not dismiss their history so easily.

48 When celebrating National Freedom Day in 1977, Nelson Mandela entered Upington Stadium driving a donkey cart

49 Jones, 'Response to Demand', p. 197.

50 Starkey, 'Donkey in South Africa'.

4 DONKEYS AND MULES AT WAR

1 Memorial inscription at the People's Dispensary for Sick Animals, Kilburn, London.

2 William Hamblin, *Warfare in the Ancient Near East to 1600 BC* (New York, 2006), p. 17.

3 Onager and donkey hybrids were widely used and highly valued and could cost more than 40 times the price of a donkey.

4 J. Roth, *The Logistics of the Roman Army at War, 260 BC–AD 235* (Leiden, 1998), p. 65.
5 Ibid., p 206. A British Service Manual of 1915 states that donkeys carry 100 lb, as compared to the mule, which can carry 160 lb and travel 20–25 miles a day, quoted in A. Dent, *Donkey: The Story of the Ass from East to West* (London, 1972), p. 165.
6 Roth, *Logistics of the Roman Army*, p. 201.
7 J. Clutton-Brock, *Horse Power: A History of the Horse and the Donkey in Human Societies* (Cambridge, MA, 1992), p. 120.
8 D. Thrapp, 'The Mules of Mars', *Quartermaster Review*: (May–June 1946), http://www.qmmuseum.lee.army.mil/WWII/mules_of_mars.htm (accessed 11 April 2009).
9 A. Castel, *Winning and Losing the Civil War* (Columbia, SC, 1996), p. 148.
10 A. Thayer Mahan, *The Story of the War in South Africa, 1899–1900*, (London, 1900).
11 There are more books available these days about the tremendous numbers of animals involved in human warfare, which are more recently commemorated with statues and plaques at war memorials and other places of remembrance. See, for example: V. G. Ambrus, *Horses in Battle* (London, 1975) and J. J. Kramer, *Animal Heroes: Military Mascots and Pets* (London, 1982).
12 This was commemorated by Rudyard Kipling in his poem 'Screw Guns'.
13 L. Travis, *The Mule* (London, 1990), p. 37.
14 E. H. Baynes, *Animal Heroes of the Great War* (London, 1925), p. 114.
15 Ibid., p. 124.
16 Ibid., p. 128.
17 A. D. Carbery, *The New Zealand Medical Service in the Great War, 1914–1918* (Auckland, 1924).
18 H. M. Alexander, *On Two Fronts: Being the Adventures of an India Mule Corps in France and Gallipoli* (London, 1917).
19 C.E.W. Bean, *The Story of ANZAC* (Sydney, 1921), vol. I, p. 394.
20 *West Australian* (20 July 1915), p. 5.

21 P. Cochrane, *Simpson and the Donkey: The Making of a Legend* (Melbourne, 1992), pp. 151–3.
22 F. M. Reed, 'Along the Burma Road', *Rutland Herald* (25 February 2008).
23 H. L. Hames, *The Mules Last Bray: World War II and US Forest Service Reminisces* (Missoula, MT, 1996).
24 O. Moore, 'Burma: Canada's Forgotten War', *Globe and Mail* (10 August 2005).
25 C. Weidenburner, 'British Raid Burma', *Life*, XXVI (1943), pp. 19–26.
26 Quoted in Travis, *The Mule*, p. 55.
27 S. McKenzie, 'The "Stealth Mules" of World War' (2008): http://news.bbc.co.uk/2/hi/uk_news/scotland/highlands_and_islands/7707748.stm (accessed 22 April 2009).
28 K. Huebner, *Long Walk through War: A Combat Doctor's Diary* (College Station, TX, 1987), p. 80.
29 Ibid., p. 154.
30 The story of the capture and training, and eventual fate, of a further 200 donkeys caught from the far north of South Australia is told in the personal notes of Captain Oswald Cundy, of the Fourth Australian Remount Squadron. It is of interest to note, however, in regard to chapter Three of this book, that the donkeys were easy to catch: 'There was a donkey team at Blinman owned and worked by one E. J. White, who was more than happy to sell them. His method was simple: the wagon with the chain traces, collars and harness lying just where they had been removed from the team, stood near the house. He sent his large tribe of barefoot children to drive them in from the common and each donkey went to its place and stood patiently while they were harnessed. They were a fair lot and, as I recall, we bought nearly all of them.' 'Reminiscing of donkeys', *Yorke Peninsula Country Times* (27 July 2007).
31 P. McLaughlan, 'Gas Guzzlers Replaced by Hay Burners: Pack Animals Help Fight War on Terror', *Veterans Magazine* (March 2005), pp. 8–11.
32 'Special Forces Use of Pack Animals' (John F. Kennedy Special

Warfare Centre and School, North Carolina, 2004).

33 McLaughlan, *Gas guzzlers*.

34 H. Trabelsi, *From Booby-Trapped Donkey to Hashish-Laden One* (2009): http://www.saudiwave.com/ (accessed 30 March 2009).

35 J. Landay, 'Election Logistics Test Officials, Donkeys', *Park City Daily News* (4 September 2005) p. 8.

5 DONKEYS IN LITERATURE, FILM AND ART

1 G. K. Chesterton, 'The Donkey' (1920): http://www.englishverse.com/poems/the_donkey (accessed 29 November 2004).

2 Apuleius, *The Golden Ass* (Oxford, 1994).

3 S. Frangoulidis, 'Lucius' Metamorphosis into an Ass as a Narrative Device', in *Witches, Isis and Narrative: Approaches to Magic in Apuleius' 'Metamorphoses'* (Berlin and New York, 2008): http://www.reference-global.com/isbn/978-3-11-020594-7 (accessed 12 May 2010).

4 This is the first listing in the OED under 'ass'; the second involves the vulgar association with the word 'arse'. In sixteenth-century English, 'ass' and 'arse' were pronounced alike – and Shakespeare enjoyed using puns. The many connotations of the word 'ass' ranged from the sacred to the profane. The sexual prowess of the ass had also continued in the representations from their association with debauched gods from the ancient world.

5 M. Cervantes, *The History of Don Quixote de la Mancha* [1605], trans. J. Ormsby (Chicago, IL, 1885).

6 Padre A. Vieira, 'Our Brother, the Donkey', in *Kinship with animals*, ed. K. Solisti and M. Tobias, updated edn (San Francisco, CA, 2006), pp. 133–8.

7 H. Kean, *Animal Rights: Political and Social Change in Britain since 1800* (London, 1998).

8 St Bonaventure, *The Soul's Journey into God: The Tree of Life: The Life of St Francis*, trans. E. Cousins (New York, 1978). Saints had often provided an alternative relationship with animals to

those presented in the fables and bestiaries. Animals served as examples of piety and revelation, of friendship and martyrdom.

9 S. T. Coleridge, 'To a Young Ass its Mother being Tethered near it' (1794): http://homepages.wmich.edu/~cooneys/poems/bad/Coleridge.ass.html (accessed 9 January 2008).

10 R. L. Stevenson, *Travels with a Donkey in the Cevennes* (London, 1988).

11 J. R. Jiménez, *Platero and I* (Austin, TX, 1956).

12 A. A. Milne, *Winnie the Pooh* (London, 1926).

13 G. Orwell, *Animal Farm* (London, 1945). Orwell uses animal fable for his political satire, the animals representing different aspects of human behaviour.

14 A. Adamson and V. Jenson, *Shrek* (2001).

15 Brabham argues that Donkey represents the 'other' in media in the form of the black stereotype as outlined by Stuart Hall: the native, the slave and the clown. Donkey supports the flow of the white fairy-tale narrative, but from behind the mask of blackness. See Daren Brabham, 'Animated Blackness in *Shrek*', *Rocky Mountain Communication Review*, III/1 (2006), pp. 64–71.

16 R. Bresson, *Au Hasard Balthazar* (1966).

17 R. Ebert, *Au Hasard Balthazar* (2004): http://rogerebert.suntimes.com/ (accessed 19 April 2007).

18 A. Merrifield, *The Wisdom of Donkeys: Finding Tranquility in a Chaotic World* (New York, 2008).

19 Ibid., pp. 244–5.

20 The most useful resources here: A. Bluhm and L. Lippincott, *Fierce Friends: Artists and Animals, 1750–1900* (London, 2005); L. Somerville, *Animals in Art* (London, 2003); and A. Dent, *Animals in Art: 116 Reproductions* (Oxford, 1976).

21 J.M.C. Toynbee, *Animals in Roman Life and Art* (London, 1973).

22 Byzantinischer Mosaizist des 5. Jahrhundert, Bodenmosaik, Szene: Kind und Esel, (Istanbul, fifth century).

23 T. Mathews, *The Clash of the Gods: A Reinterpretation of Early Christian Art* (Princeton, NJ, 1993), p. 48.

24 L. Hobgood-Oster, *Holy Dogs and Asses: Animals in the Christian Tradition* (Chicago, IL, 2008).

25 The ox and ass, although not mentioned in the synoptic gospels, became all pervasive in representations of the *Nativity* and *Adoration of the Magi*. Many religious paintings derive from the apocryphal Gospels, the iconography arising from Isaiah (1:3): 'The ox knoweth his owner, and the ass his master's crib: but Israel hath not known me, and my people hath not understood'. This text has led to the understanding that the ox represents the Jews, who do not accept Jesus, and the donkey the Gentiles, or pagans, who do.

26 *Donkey* from *Animals after the First Masters for Examples in Drawing, Engraved under the Superintendance of George Cooke, after Paulus Potter* (London, 1 July 1829).

27 Susanna Partsch, *Franz Marc, 1880–1916* (Cologne, 2006).

28 J. Batty, *Landseer's Animal Illustrations* (Hampshire, 1990).

29 S. Bitton and E. Sanbar, *Mahmoud Darwish: As the Land is Language* (Israel/France, 1997).

Select Bibliography

A History of the Beltana Pastoral Company Ltd (Adelaide, 1965)

Adamson, A. and V. Jenson, 'Shrek' (2001)

Aganga, A. A., et al., 'Carcass Analysis and Meat Composition of the Donkey', *Pakistan Journal of Nutrition*, II/3 (2003)

Alexander, H. M. Major, *On Two Fronts: Being the Adventures of an India Mule Corps in France and Gallipoli* (London, 1917)

Apuleius, *The Golden Ass*, English edn (Oxford, 1994)

Baker, S. W., *Exploration of the Nile Tributaries of Abyssinia* (Hartford, 1868)

Baker-Carr, J., *An Extravagance of Donkeys* (Colorado, IL, 2006)

Barker, H. M., *Camels and the Outback* (Carlisle, 1995)

Batty, J., *Landseer's Animal Illustrations* (Hampshire, 1990)

Baynes, E. H., *Animal Heroes of the Great War* (London, 1925)

Bean, C.E.W., *The Story of ANZAC* (Sydney, 1921)

Beja-Pereira, A., et al., 'African Origins of the Domestic Donkey', *Science*, CCCIV (2004)

Bitton, S., and E. Sanbar, *Mahmoud Darwish: As the Land in Language* (Paris, 1997)

Bluhm, A., and L. Lippincott, *Fierce Friends: Artists and Animals, 1750–1900* (London, 2005)

Borwick, R., *The Book of the Donkey* (London, 1981)

Bough, J., 'Value to Vermin: The Donkey in Australia', unpublished thesis, the University of Newcastle (Newcastle, 2008)

——, 'The Mirror has Two Faces: Contradictory Reflections of Donkeys in Western Literature from Lucius to Balthazar',

Animals, I/56–68 (2011)
Bresson, R., 'Au Hassard Balthazar' (1996)
Brookshier, F., *The Burro* (Norman, 1974)
Bulliet, R., *Hunters, Herders and Hamburgers: The Past and Future of Human-animal Relationships* (New York, 2005)
Burbidge, K., *The Donkey Drivers* (Hermitage, 1986)
Carbery, A. D., *The New Zealand Medical Service in the Great War, 1914–1918* (Auckland, 1924)
Castel, A., *Winning and Losing the Civil War* (Columbia, IN, 1996)
Cervantes, Miguel de, *The History of Don Quixote de la Mancha* [1605], trans., J. Ormsby (Chicago, IL, 1885)
Clutton-Brock, J., *Horse Power: A History of the Horse and the Donkey in Human Societies* (Cambridge, MA, 1992)
Cochrane, P., *Simpson and the Donkey: The Making of a Legend* (Carlton, 1992)
Dent, A., *Donkey: The Story of the Ass from East to West* (London, 1972),
Fielding, D., and P. Krause, *Donkeys* (Oxford, 1998)
——, and R. A. Pearson, eds, *Donkeys, Mules and Horses in Tropical Agricultural Development* (Edinburgh, 1991)
——, and P. Starkey, eds, *Donkeys, People and Development: A Resource Book of the Animal Traction Network for Eastern and Southern Africa (ATNESA)* (Wageningen, 2004)
Flannery, T., *The Future Eaters: An Ecological History of the Australian Lands and Ppeople* (Melbourne, 1994)
Frangoulidis, S., 'Lucius' Metamorphosis into an Ass as a Narrative Device', in *Witches, Isis and Nnarrative: Approaches to Magic in Apuleius' "Metamorphoses"* (Berlin and New York, 2008)
Fudge, E., *Perceiving Animals: Humans and Beasts in Early Modern English Culture* (Urbana, IL, 2002)
Griffith, M., 'Horse and Donkey Work: Equids and the Ancient Greek Imagination', *Classical Philology*, CI/3 (2006)
Hamblin, W., *Warfare in the Ancient Near East to 1600 BC* (New York, 2006)
Hames, H. L., *The Mules Last Bray: World War II and US Forest Service Reminisces* (Missoula, MT, 1996)

Herodotus, *Herodotus: The Histories*, trans., R. Wakefield (Oxford, 1998)
Hobgood-Oster, L., *Holy Dogs and Asses:Animals in the Christian Tradition* (Urbana and Chicago, IL, 2008)
Huebner, K., *Long Walk through War: A Combat Doctor's Diary* (College Station, TX, 1987)
Jacobs, N., 'The Great Bophuthatswana Donkey Massacre: Discourse on the Ass and Politics of Class and Grass', *American Historical Review*, CIX (2001)
——, *Environment, Power and Injustice* (Cambridge, 2003)
Jimenez, J. R., *Platero and I* (Austin, TX, 1956)
Kean, H., *Animal Rights: Political and Social Change in Britain since 1800* (London, 1998)
Kennedy, M. J., *Hauling the Loads: A History of Australia's Working Horses and Bullocks* (Melbourne, 1992)
Kimura, B., et al., 'Ancient DNA from Nubian and Somali Wild Ass provides insight into Donkey Ancestry and Domestication', *Proceedings of the Royal Society* (28 July 2010): rspb.royalsociety-publishing.org (accessed 1 February 2011)
Kramer, J. J., *Animal Heroes: Military Mascots and Pets* (London, 1982)
Kugler, W., H-P. Grunenfelder, and E. Broxham, *Donkey Breeds in Europe: Inventory, Description, Need for Action, Conservation* (Basel, 2008)
Long, W., *Asses vs Jackasses* (Portland, OR, 1969)
Lummis, C., 'Brother Burro', *Land of Sunshine*, IV/3 (1896)
——, *Flowers of our Lost Romance* (New York, 1929)
Mahaffy, P. J., 'On the Introduction of the Ass as a Beast of Burden into Ireland', in *Proceedings of the Royal Irish Academy*, (London, 1917)
Malamud, R., *Poetic Animals and Animal Souls* (New York, 2003)
Mallam, P., *The Donkey Book* (London, 1935)
Manning, A., and J. Serpell, eds, *Animals and Human Society* (London, 1994)
Marshall, F., *Ethnoarchaeological Perspectives on the Domestication of the Donkey: Bone Commons Forum* (2006): www.alexandria archive.org/icaz/icazForum/ (accessed 1 January 2008)

——, 'How Wild Asses became Donkeys to the Pharaohs', *New Scientist* (22 March 2008): www.newscientist.com/channel/life/mg19726474.700-how-wild-asses-became-donkeys-of-the-pharaohs.html

Mason, I. L., *A World Dictionary of Livestock Breeds, Types and Varieties*, 3rd edn (Wallingford, Oxon, 1988)

Mathews, T., *The Clash of the Gods: A Reinterpretation of Early Christian Art*, (Princeton, NJ,1993)

Mayhew, H., *London Labour and the London Poor* (London, 1851)

McCandless, S. M. *The Burro Book* (Pueblo, CO, 1900)

Meadow, R., and H.-P. Uerpmann, eds, *Equids in the Ancient World* (Wiesbaden, 1986)

Merrifield, A., *The Wisdom of Donkeys: Finding Tranquility in a Chaotic World* (New York, 2008)

Milne, A. A., *Winnie the Pooh* (London, 1926)

Moehlman, P. D., F. Kebede and H. Yohannes, 'The African Wild Ass (*Equus africanus*): Conservation Status in the Horn of Africa', *Applied Animal Behaviour Science*, LX/2–3 (1998), pp. 115–24

Morrell, V., 'Cruelest Place on Earth: Africa's Danakil Desert', *National Geographic* (October 2005)

Olsen, S., ed., *Horses Through Time* (Boulder, CO, 1996)

Orwell, G., *Animal Farm: A Fairy Story* (London, 1951)

Ritvo, H., *The Animal Estate: The English and Other Creatures in the Victorian Age* (Cambridge, MA, 1987)

Rossel, S., et al., 'Domestication of the Donkey: Timing, Processes and Indicators', *PNAS: Proceedings of the National Academy of Sciences of the USA*, CV/10 (2008)

Roth, J., *The Logistics of the Roman Army at War (260 BC – AD 235)* (Leiden, 1998)

Sindawi, K., 'The Donkey of the Prophet in Shi'ite Tradition', *Al-Masaq*, XVIII/1 (2006)

Solisti, K., and M. Tobias, eds, *Kinship with Animals* (San Francisco, CA, 2006)

Somerville, L., *Animals in Art* (London, 2003)

Stamm, M., *The Mule Alternative* (Battle Mountain, NV, 1992)

Starkey, P., ed., 'The Donkey in South Africa: Myths and Misconceptions', in *Animal Traction in South Africa: Empowering Rural Communities* (Gauteng, 1995)
Stevenson, R. L., *Travels with a Donkey in the Cevennes* (London, 1988)
Svendsen, E., ed., *The Professional Handbook of the Donkey* (London, 1986)
Swinfen, A., *The Irish Donkey* (Dublin, 2004)
Tegetmeier, W. B., and C. L. Sutherland, *Horses, Asses, Zebras, Mules and Mule Breeding* (London, 1895)
Thackeray, W., 'Respecting Asses', *Cornhill Magazine* (January–June 1864)
Thrapp, D., 'The Mules of Mars', *The Quartermaster Review* (May–June 1946)
Tobias, M., and J. Morrison, *Donkey: The Mystique of Equus Asinas* (San Francisco, CA, 2008)
Toynbee, J.M.C., *Animals in Roman Life and Art* (London, 1973)
Travis, L., *The Mule* (London, 1990)
Tristram, H. B., *The Natural History of the Bible* (London, 1889)
Vandenbeld, J., *Nature of Australia: A Portrait of the Island Continent* (Sydney, 1988)
Vernon, A., *The History and Romance of the Horse* (New York, 1946)
Walker, A., *Australian Donkeys* (Victoria, 1973)
Young, R., *Analytical Concordance to the Bible* (Grand Rapids, MI, 1982)
Zeuner, F. E., *A History of Domesticated Animals* (London, 1963)

Associations and Websites

There are numerous societies, organizations and sanctuaries around the world dedicated to donkeys and mules, their breeding, care, use and welfare. Many of these can be found at:

Donkey Association/Organizations Regional Donkey Clubs
www.equineseek.com › Donkeys/Mules

A selection of those organizations follows:

The Brooke
www.thebrooke.org

Donkey Breed Society
www.donkeybreedsociety.co.uk/

The Donkey All Breeds Society of Australia
donkeyallbreedsaustralia.org/

The British Mule Society
www.britishmulesociety.co.uk/

The American Donkey and Mule Society
www.lovelongears.com/

The Animal Traction Network for Eastern and Southern Africa (ATNESA)
www.atnesa.org/

Donkey and Mule Society of New Zealand
www.donkey-mule.org.nz/

Miniature Mediterranean Donkey Association
www.miniature-donkey-assoc.com/links.htm

Federation nationale anes et randonnees (FNAR)
www.ane-et-rando.com/

Association Nacional de Criadores de la Raza Asnal Andaluzu (ANCRAA)
www.ancraa.org/

Canadian Donkey and Mule Association
www.clrc.ca/donkey.shtml

DONKEY SANCTUARIES:

The Donkey Sanctuary, Devon, UK
http://drupal.thedonkeysanctuary.org.uk

Other donkey sanctuaries around the world are listed on the National Miniature Donkey Association web page:

www.nmdaasset.com/sanctuaries.php

A few examples include:

Aruba
www.arubandonkey.org/

Australia
www.donkeyrescue.org.au/
www.donkeywelfare.com.au/

Bonaire
www.donkeysanctuary.org/

Canada
www.thedonkeysanctuary.ca/website/index.php

Corfu
www.corfu-donkeys.com/

Cyprus
www.windowoncyprus.com/donkeys.htm

Israel
www.safehaven4donkeys.org/

Spain
www.nerjadonkeysanctuary.com/

USA
www.donkeyrescue.org
www.wldburrorescue.org
www.longhopes.org

Acknowledgements

I would like to thank family, friends and colleagues for their encouragement throughout the process of writing this book – and for listening to endless hours of stories about donkeys! Special thanks go to my husband for his help, support and patience, and to my mother from whom I inherited a love of donkeys. Special thanks also go to Wendy Michaels and Marea Mitchell who read drafts of the book. I am grateful to those who willingly shared their photographs with me; to Jenny Marriott and Owen Michaels-Hardy who generously lent their computer skills; and to staff at Reaktion who helped to guide me, especially Harry Gilonis for leading me through the mysteries of illustrations. I would especially like to acknowledge all those donkeys that continue to inspire me as they accompany and serve humans everywhere; and all those humans who work tirelessly to alleviate the suffering of donkeys worldwide.

Photo Acknowledgements

The author and publishers wish to express their thanks to the below sources of illustrative material and/or permission to reproduce it. (Some sources uncredited in the captions for reasons of brevity are also given below.)

The Art Institute of Chicago: pp. 50, 70; photos courtesy the Australian War Memorial, Campbell, ACT: pp. 112 (ref. B01618), 115 (ref. E00091), 124 (ref. 056326); courtesy of the author: p. 29; photo Brad B. Biggers: p. 104; Bodleian Library, Oxford: pp. 8, 22, 25 (top right), 113, 133; Boston Museum of Fine Arts: p. 67; from Sebastian Brant, *Das Narren Schyff. Ad Narragoniam* (Basel, 1494): p. 131; British Museum, London (Department of Prints and Drawings): pp. 55, 59, 110; photos © Trustees of the British Museum: pp. 55, 59, 110, 132; photo Kimberly Brown-Azzarello: p. 24; from Buffon's *Histoire Naturelle, Générale et Particulière*, vol. XXII (Paris, 1799/1800): pp. 14, 42; The Cleveland Museum of Art: p. 69; photo Giovanni Dall'Orto: p. 66; photo Mar Daniel/Dreamstime.com: p. 34; photo Debu55y/Dreamstime.com: p. 57; Egyptian Museum Berlin: p. 46 (top); photo Ed Emery: p. 31; photo © Fleyeing/2011 iStock International Inc.: p. 9 (foot); The Frick Collection, New York: p. 155; photo Tony Fricko: p. 9 (top); photo David Haberlah: p. 16; The Armand Hammer Museum of Art and Cultural Centre, Los Angeles: p. 137; photo Shara Henderson/Mood Board/ Rex Features: p. 6; photo Martha Howard, Elms Farm Miniature Donkeys, Mason, OH (www.donkeys.net): p. 35; Library of Congress, Washington, DC (Prints and Photographs Division): pp. 15, 79, 81, 82,

85, 109, 120; (Prints and Photographs Division – Morgan Collection of Civil War drawings): p. 111; photo © Lingbeek/2011 iStock International Inc.: p. 100; from S. M. McCandless, *The Burro Book* (Pueblo, co, 1900): p. 80; photo © Charlotte McGuire/2011 iStock International Inc.: p. 138; photo Servel Miller: p. 38; morguefile.com: p. 36; Musée des Beaux-Arts de la Ville du Paris: p. 156; Musée Cemusci, Paris: p. 51; Musée Cognacq-Jay, Paris: p. 71; Musée du Louvre, Paris: pp. 62, 68, 72, 149; from Eadweard Muybridge, *Animal Locomotion: an Electro-Photographic Investigation of Consecutive Phases of Animal Movements 1872–1885* (Philadelphia, 1887): p. 130; photo Nancy: p. 95; National Library of Australia, Canberra: p. 88; Northern Territory Library, Darwin, NT: p. 91; from John Ogilby, *Fables of Aesop Paraphras'd in Verse: Adorn'd with Sculptures and Illustrated with Annotations* (London, 1665): p. 132; photo © parema/2011 iStock International Inc.: p. 135; private collection: p. 162; photo Frank M. Rafik: p. 46 (top); photo Redwood202/Dreamstime.com: p. 41; photos © Roger-Viollet: pp. 51, 63, 127; photo Dave Sag: p. 19; photo JM Schneid: p. 86; Scrovegni Chapel, Padua: pp. 151, 153, 154; photo © Tony Smith (www.taffysmith.com): p. 30; Staatliche Kunstsammlungen Dresden: p. 152; from R. L. Stevenson, *Travels with a Donkey in the Cevennes* (London, 1879): p. 140; Stoa of Attalus Museum, Athens: p. 66; photo sure2talk: p. 58; Ta-t'ung City Institute of Archaeology, Shan-hsi, China: p. 49; from W. B. Tegetmeier and C. L. Sutherland, *Horses, Asses, Zebras, Mules and Mule Breeding* (London, 1895): pp. 23, 25 (foot), 32; Alexander Turnbull Library, National Library of New Zealand, Wellington: 119; Palazzo Medici-Riccardi, Florence: p. 61; Walker Art Gallery, Liverpool (National Museums and Galleries on Merseyside): p. 157; Werner Forman Archive/British Museum, London: p. 107; Werner Forman Archive/Egyptian Museum, Turin: p. 46 (foot); Werner Forman Archive/Rothschild Wine Museum, Bordeaux: p. 52; photo © WLDavies/2011 iStock International Inc.: p. 11; photo zeddy 1200: p. 37.

Index